MW01388117

Satellite Communications Pocket Book

Satellite Communications Pocket Book

Revised Edition

Eur Ing, James Wood, CEng, FIEE

Newnes
An imprint of Butterworth-Heinemann
Linacre House, Jordan Hill, Oxford OX2 8DP
A division of Reed Educational and
Professional Publishing Ltd

A member of the Reed Elsevier plc group

OXFORD BOSTON JOHANNESBURG
MELBOURNE NEW DELHI SINGAPORE

First published 1994
Revised edition 1996

British Library Cataloguing in Publication Data
Wood, James
Satellite Communications Pocket Book
I. Title
621.382

ISBN 0 7506 1749 7

Library of Congress Cataloguing in Publication Data
Wood, James,
Satellite communications pocket book/James Wood.
p. cm.
Includes bibliographical references and index.
ISBN 0 7506 1749 7
1. Artificial satellites in telecommunication. I. Title.
TK5104.W57
384.5'1-dc20 93-39894
CIP

Printed in England by Clays Ltd, St Ives plc

Contents

Preface

Technology is moving ahead at a breathless pace. Nowhere is this more in evidence than in satellite communications and television. In the next three years a communications revolution will sweep across America. It will be fuelled by the most powerful advances ever made since the birth of the satellite age; the merging of three stunning technologies, high power direct broadcast satellites, digital video compression and fibre optics.

For the first time in 20 years here is a technological revolution not made in Japan. From a late start in introducing satellite television, America has bypassed Japanese and European technology and it is now a virtual certainty that these two regions of the world will adopt US technology and standards.

This book though primarily dedicated to satellite communications describes development being made in related technologies which will eventually lead to better global communications and high definition television broadcast from satellites and over existing terrestrial UHF-TV transmitters.

The book is written in a style which makes informative, yet easy reading and an invaluable source of information for communications engineers, students, enthusiasts and those working in the satellite TV industry.

Preface

Acknowledgements

Acknowledgement is made to the following for the use of material:

- *Satellite companies* Alpha Lyracom; Astra; Anik; Asiasat; Insat; AT&T; Italsat; Arabsat; Brazilsat; British Telecom; Brightstar; Eutelsat; France Telecom; GE Americom; GTE Spacenet; Hughes Communications Inc.; Inmarsat; Intelsat; Loral Corp.; Morelos; Optus Communications; Palapa.
- *Manufacturers* Aerospatial; Alcatel Espace; Arianespace; Andrew Corp.; ADC; Echosphere Corp.; General Dynamics; GE AstroSpace; Harris Corporation; Hughes Aircraft Company; Hughes Space & Communications; Hughes Network Systems; Intelesys; Marconi GEC; Matra-Marconi; Martin Marietta; Motorola Corp.; McDonnell-Douglas; Mobile Telesystems Inc.; Orbitron Corp.; Radiation Systems; Scientific Atlanta; Space Systems/Loral; Thomson Consumer Electronics; Thomson Tubes Electroniques; Philips Consumer Electronics; ViaSat Technology Corp.; Varian.
- *News agencies and broadcasting companies* ABC; BBC; BBC-WSTV; CBS; CNN; NBC; IBA; ITN; Reuters; SISlink; Star-TV; Telecom Satellite Corporation Japan; NHK; Worldnet; WTN; VisNews.

For the use of copyright material the author would like to acknowledge: EMAP Business Publishing (*International Broadcasting*); American Radio Relay League (*ARRL Handbook*); Butterworth-Heinemann; Communications Technology Corp. (*International Cable*); Institution of Electrical Engineers (IEE); Phillips Business Information Systems (*Via Satellite*); Salamandar Books.

Personal thanks go to: Scott Chase, Editor of *Via Satellite*; Wayne H. Lasley, Editor of *International Cable*; Margaret Riley, Commissioning Editor (Focal Press), Duncan Enright, Commissioning Editor and Alison Boyd, Desk Editor, all at Butterworth-Heinemann.

1 Introduction

There have been four separate ages in communications and broadcasting. These have been defined as:

(a) the (first) wire age 1840–1900, the undersea telegraph
(b) the wireless telegraphy age, the longwave telegraph transmitter
(c) the age of sound broadcasting for entertainment purposes, 1920
(d) the age of television broadcasting, 1939–1946 to the present day

To these can now be added the satellite age, which began in 1965, and the age of fibre optic communications, which has already begun. With the exception of the latter, now in its early stages, each of the other ages has earned its place in history, and each in its own way has been responsible for creating a new market in communications and broadcasting, thereby stimulating an increased global demand.

The coming of television had nothing like the same impact as sound broadcasting in the AM wavebands. The reasons were both economic and technical. Television sets had small screens but at the same time were relatively expensive. Technical limitations in line-of-sight transmission by VHF meant that, unlike radio broadcasting in the AM waveband, where a single radio station could be heard over long distances, television broadcasting called for the construction of many television stations. For most countries this meant that the capital city was served by television but the provinces were neglected. Television broadcasts were in monochrome with only moderate picture quality.

The introduction of 525-line transmissions in America and 625-line transmissions in Europe, coupled with the availability of cheaper sets, and the building of more television stations, increased the popularity of television but it was really the coming of colour that set the seal on its potential. Even so, no one could have possibly foreseen that the time would come when television would become the most important form of mass medium (on a national basis as far as Great Britain was concerned, since there was no way in which VHF-TV signals could be transmitted or received from the continent of Europe, much less America).

In 1945, only a few months after the end of World War II, an article appeared in *Wireless World* written by the then little-known writer Arthur C. Clarke. In it he laid out a virtual blueprint for a new age in radio and television broadcasting and communications.

Clarke in his article went on to write what has come to be accepted as the most remarkable prophecy of the twentieth century. This was his detailed hypothesis that three such satellites spaced exactly 120° apart, and travelling from west to east above the equator in geosynchronous orbit, could provide an international radio and television communication system. What this meant exactly was that a radio or a television signal could be sent to the other side of the world simply by a satellite relay of three uplinks and three downlinks in an alternating sequence.

Clarke's thinking was far ahead of any existing technology at that time but in 1957 the CIS (then USSR) launched Sputnik 1, a satellite that circled the earth every 90 minutes. Though primitive in design compared to the satellites that were developed several years later, it signalled the beginning of the satellite age. In 1963 the Americans launched the first geosynchronous version fitted with active transponders.

In 1965 the USA transmitted a colour picture of an unfurled US flag to Europe, where the transmission was received and relayed on national television networks. In order to transmit a single television transmission over the satellite it was necessary to shut down all 270 speech channels.

The next step forward was in more sophisticated designs of satellites able to handle 1–3 television channels along with a few thousand voice and data channels. In order to appreciate more fully the impact that the satellite age had, one need only look at the earth itself. Some of the regions of the world are made of harsh geographical features, mountains, rivers, valleys and ravines, and, in the Middle East particularly, barren and hostile deserts. Before the coming of the communications satellite, millions of people in remote areas of Third World countries had never seen a television set. For these people the only means of contact with the outside world was the ubiquitous battery-operated, first-generation type transistor receiver mass-produced by Japan in the early 1960s. The coming of communication satellites made it possible for Arab states to invest in downlink earth stations able to pick up television programmes from the studios in the capital city and thence redistribute these same programmes over a cable network in small towns. Arab countries were some of the first to capitalize on the advantages offered by modern technology.

The satellite age ushered in by the creation of Intelsat, Eutelsat and Arabsat, was soon followed by the second wave. This came from the well-established international communications companies like AT&T which had gone from nineteenth-century telegraphy to wireless, then shortwave and now satellite communications. Others followed, RCA, Hughes, Western Union and a few others seeking to expand into this new medium. Each launched its own fleet of the newer generation of satellites, and each acquired its own name. Satellites collectively became known as 'birds', with imaginative names like Spacenet, Westar, Telstar, Satcom, Galaxy, etc.

The commercial satellite age had begun. Owners switched from selling traffic by the mile, as was the custom with wireless telegraphy, to selling satellite capacity by time in minutes, hours, weeks, or even by the year – the longer the unit, the cheaper the leasing cost. Selling satellite use by time rather than distance is logical, since every satellite transmission entails a round distance of 70 000 km on the uplink and downlink even if the horizontal coverage is a few hundred kilometres.

The next step forward, the use of satellites to broadcast direct to homes, was slower in coming because of the limitations in technology. Firstly, any DBS system called for higher output powers than those currently used in communication satellites. Secondly, a satellite receiver for the mass domestic market had to be produced at a cost within reach of an average person. At the same time this must be achieved without any significant drop in technical performance. The technical criterion against which any DBS system is judged is that the quality of television reception must be at least equal to the pictures received by terrestrial transmitting systems.

The advent of satellite broadcasting on a direct-to-home basis (DTH) in 1988, pioneered by Japan, brought about the possibility of employing a much wider bandwidth of 27 MHz compared with the 6 MHz bandwidth available with UHF-TV broadcasting. Thus, the introduction of DBS in Japan provided the gateway for the introduction of high-definition television (HDTV), using 1050/60 lines in place of the existing NTSC standard 525/60.

The USA has lagged behind Japan and Europe in the introduction of direct broadcasting to viewers by satellite, but the reason for this delay is not that it lacked the technological resources, but rather the contrary. The USA has built up an infrastructure of delivery systems consisting of VHF, UHF, MMDS, LPTV, and combinations of these with cable networks. The result is that most US citizens have for many years enjoyed a wide choice of television programmes ranging from 25 to as many as 90 different channels. Against this backdrop of choice, for any DBS system to survive commercially it must be able to offer the same choice of viewing.

The extraordinary growth that has taken place in the terrestrial and now satellite television industry would not have been possible without the remarkable developments in components like transistors, semiconductors and integrated circuits. From the development of the first transistor in 1947 and the first integrated circuit a few years later there has grown a truly massive industry.

Satellite television has experienced major growth since it began in 1983 and it is expected to continue to grow. The question is, how quickly? As satellite technology improves, with more powerful versions being launched, this is bound to increase the popularity of satellite television. However, with the present permissible spatial separation of 2° between adjacent satellites, the equatorial orbital path can only accommodate so many satellites, so a distinct possibility is that other types of orbit may be considered.

Alternatively, there is another communications technology, fibre optic systems – the sixth age in communications and broadcasting. To build a number of fibre optic systems around the world as main trunk systems for speech, data and television is entirely feasible. This is how the world's first communication system was built in the nineteenth century, except that the cables contained metallic pairs of conductors. The use of fibre optic cables for carrying speech and television signals is not a new, untried technology.

The use of cables for communications is far older than that of wireless and satellite, and at first it was thought that the introduction of satellites would spell the end for cables. This thinking has now been reversed; the reality is that, after 25 years of satellite communications, both coaxial and fibre optics are expanding. Optical systems offer greater bandwidth and possess a greater immunity to the difficulties experienced with terrestrial and satellite communications, such as electrical interference and jamming.

By the year 2010 it is possible that the average household will have a combination of delivery systems, terrestrial UHF/satellite/coaxial and fibre optics, with interactive capabilities. Most television viewers of today are generally satisfied with the picture quality on their television set, because they have not experienced anything better. Yet the truth is that even the colour televisions of today cannot match the picture quality as seen on cinema screens. The reason why viewers of television are satisfied is that the small screen masks the poor resolution that is present, which would be obvious on a large screen size of 30 inches.

There are other defects with present television transmission standards, giving rise to cross-luminance and chrominance. This is why we need to see the introduction of HDTV on a 1250/50 format instead of the existing 625/50 in Europe. High definition is a relative term; even when 1250/50 HDTV is introduced, it will not be the ultimate, but rather one more step forward in the pursuit of excellence. Once viewers have seen television on a wide screen size of 36 inches in HDTV and free of unwanted artifacts no one will want to go back to present television standards.

At one time it was thought that the only path for HDTV was by satellite transmission because of the wider bandwidth it offered – 27 MHz against terrestrial UHF with its 6 MHz. This is no longer the case. Developments in digital compression have made it possible to transmit HDTV over existing UHF and cable delivery systems, so now there are no obstacles in the way of HDTV.

2 Historical overview of communications and radio broadcasting

Evolution of media

Every new medium of information has made advances on the previous generation of technology and in so doing has established new values, created an awareness of increased potential, and thereby stimulated a greater demand. The first cable age of the 1850s was no exception to this rule, but it was the introduction of radio broadcasting in the 1920s, its subsequent growth years from the 1930s and the introduction of television broadcasting in Europe and North America that eclipsed all other forms of media.

The age of television broadcasting

Experimental broadcasting with television began in the USA and Europe about the same time, 1927–1929. The USA was the first country to make an experimental transmission, carried out by AT&T in New York City. Two years later John Logie Baird, the lone British inventor, made an experimental broadcast for the BBC. Other experiments were proceeding in France and Germany.

But even before the late 1920s, investigations into transmitting pictures had been going on. The first to achieve any success was Paul Nipkow with his scanning disc. Holes in the scanning disc permitted some light reflected from a screen to hit a phototube where a tiny current flow was generated in proportion to the amount of light falling onto it.

Not all of the ideas had a future; some were impractical, whilst others were incapable of transmitting anything other than a low-definition picture. In this context it should be noted that low and high definition are relative terms. In the late 1920s a 100-line picture would have been regarded as high definition, since most of the early experiments were with 25- and 50-line pictures.

Although Logie Baird is credited by the public as being the first in Great Britain to develop an experimental television transmission, his efforts were eventually doomed to failure because of his inability or reluctance to try other systems. By comparison, the research and development effort being expended by Marconi EMI Television was huge. But even this was overshadowed by the research investment being made by American companies.

The first regular television broadcasts took place from the BBC using the EMI system. Also in 1939 television broadcasts were made from the World Fair in New York. However, in both countries any further work on television broadcasting was set back because of the war. It got under way in 1946, by which time regular broadcasting was taking place in the USA. The first television sets in Britain were of the small-screen variety, and transmissions were at first limited to the London area.

Yet even at this time thought was being given to the transmission of colour pictures. A number of systems were proposed but the problem was one of compatibility with monochrome. That is, it was considered essential that colour transmissions could still be seen satisfactorily, albeit in monochrome, with existing television receivers.

In 1963 the USA standardized on a system decided by the National Television Standards Committee which became known as NTSC. It used a 525-line 60-field system. Four years later Europe brought out two more television standards for colour television. These were PAL and SECAM. PAL is a 625-line 50-field system; SECAM is also a 625-line 50-field system but with other differences to do with modulation characteristics. To convert from one standard to another required the employment of a standards converter.

Emergent technologies

Telecommunications authorities, broadcasting authorities and the communications satellite companies are spending millions of dollars investing in new technologies and the last decade has seen an unprecedented level of expenditure. These technologies include microwave communications, sound broadcasting, UHF television broadcasting, super power UHF in the USA, new video recording technologies, electronic news gathering (ENG), delivery of television programmes through cable-satellite delivery systems (SMATV), and multi-point, microwave delivery systems (MMDS). In recent years direct broadcast by satellite (DBS) has taken off in Europe and in Japan. To all these various means of delivering television to millions of viewers has been added the prospect of global delivery by fibre optics. After the first wire age of the nineteenth century, and the second wire age of the twentieth century which linked the USA and Europe by transatlantic coaxial cable, the twenty-first century promises the age of fibre optic systems.

The influence of satellites on cable television

Although cable distribution systems have been around for a long time, with a history older than that of wireless, it was the coming of the communications satellite with its ability to transmit high-quality television pictures from one continent to another, and from one part of the USA to another, which influenced the growth of cable television. From a master satellite receiving system, thousands of viewers could be connected by coaxial cable, giving these subscribers the choice of many different programmes.

In the USA alone cable penetration has gone up from a mere 11% in 1973 to 53% by 1987. If this growth rate is sustained – and there is no reason to think otherwise – then by 1995 cable television could dominate other distribution systems, with a penetration of 75%. It is for this reason that broadcasters in the US are about to start up DBS.

Some milestones in satellites, video, and HDTV technologies

1945 Arthur C. Clarke makes his prophecy on satellite communications.
1956 First ever videotape transmission, pictures from CBS New York to CBS Television City, Hollywood.
1957 The CIS (then USSR) makes history by launching Sputnik 1 into space.
1960 The USA launches two experimental satellites, Echo 1 and Echo 2.
1965 The USA launches the first geosynchronous satellite, Early Bird (Intelsat I).
1965 First ever video telecast between the CIS (then USSR) and the USA.
1967 The USA launches a second geosynchronous satellite, Intelsat II.
1969 The USA launches a third geosynchronous satellite, Intelsat III.
1970 The three Intelsat satellites, spaced 120° apart, provide global communications; Clarke's prophecy is realized.
1978 NHK Japan initiates HDTV experimental transmissions via satellite.
1981 The USA launches the first space shuttle.
1982 Two HDTV systems are demonstrated at the IBC, Brighton, UK.
1985 NHK Japan initiates regular transmissions of DBS.

The role of satellite television in today's society

Television's relationship with society has always been a topical issue, but in recent years the role of television has been elevated to a more powerful position. Conceived in the beginning as a means of broadcasting visual entertainment, television has now assumed a status unequalled by any other branch of the media. Information broadcasting in its broadest sense encompasses propaganda broadcasting to induce people to buy goods, and news, designed to truly inform or to influence and even deceive.

More effectively than the purely spoken or printed word, television, with its visual impact on viewers, is capable of exerting considerable influence on a captive audience. This influence directed to a mass audience can be either positive or negative, depending on who controls it, and whose interest it expresses. Television's role in modern society is therefore an expanding one.

In the context of satellite broadcasting making it easier for viewers to be influenced by broadcasts from another country, it is as well to bear in mind the essential differences between the traditional satellite delivery systems and DBS methods. Traditional satellite television broadcasting takes a fairly large receiving system with big dishes to access a satellite system. The received television signal is then relayed over the country's usual UHF terrestrial television delivery network. Alternatively, the foreign programmes are distributed over satellite–cable television networks. So, if the state-controlled television network or the satellite–cable company do not feel it is in viewers' – or their – interest that the foreign programmes are broadcast, then the foreign programme may not be screened.

Direct satellite broadcasting operates on a completely different level; here the foreign programmes are picked up on a DTH basis and there is no third party intervention. DBS for this reason may become the most powerful form of media of all time, because it cannot be censored.

3 The satellite age

The satellite

In the December 1945 issue of *Wireless World*, Arthur C. Clarke wrote:

> An artificial satellite at the correct distance from the earth would make one revolution every 24 hours, i.e., it would remain stationary above the same spot and would be within optical range of nearly half of the Earth's surface. Three repeater stations, 120 degrees apart in the correct orbit could give television and microwave coverage to the entire planet.

In theory all things are possible; it is only when engineers and scientists attempt to put theoretical ideas into practice that difficulties become evident. This is a fact of life and is not intended to detract from the importance of Arthur C. Clarke's vision. In order to launch a satellite into space, a launch vehicle is needed, and also exact navigation, to say nothing of all the other problems, such as rocket fuel and electrical power for the satellite's communication system. At that time in 1945 none of these problems had been tackled by research, and nor did the National Aeronautics and Space Administration exist. Yet within the timescale of another 20 years, Clarke's vision became a reality.

In December 1947 Dr William Shockley of Bell Telephone Labs changed the course of history by demonstrating to his colleagues what he called the *transistor effect*. From this demonstration, and a later one on 30 June 1948, sprang one of the most important inventions of the twentieth century. The transistor represented a real alternative to the thermionic vacuum tube. Tubes were big and fragile, and had a filament which consumed electrical power, and therefore possessed a wear-out mechanism, which gave them a limited life. Because a transistor had no wear-out mechanism, in theory it had an infinite life. Subsequent experience showed that the transistor had a tendency to sudden catastrophic failure when subjected to heat or voltage transients.

The transistor went on to replace the thermionic vacuum tube in many sectors, but more than this, the possibility of building many transistors on a single chip gave way to the integrated circuit (IC). ICs made it possible to build a new generation of equipment, and a new industry was born, 'electronics'. The IC opened the gateway to a new world of computers, microprocessors, imaging devices, and ultimately very powerful computers needed for satellite systems.

In 1957 the CIS (then USSR) launched Sputnik 1. It was a low-orbit satellite broadcasting radio signals to the earth on 31.5 MHz. As present satellite technology goes, Sputnik 1 was a primitive piece of hardware, but it was, notwithstanding, a tremendous achievement, the first milestone in man's quest for domination of space.

Not only was the launch successful but so was the spacecraft itself, right on schedule for a time of 90 days before burning up. Sputnik 1 orbited the earth 16 times every 24 hours, during which time it sent its signals back to earth from its radio transmitter.

That same year the CIS launched a further two satellites; these, too, completed all their missions without mishap. At that time the USA was totally unaware of the Soviet advances in rocket and spacecraft technology.

In 1960 the USA launched two satellites: Echo 1 and Echo 2. These satellites were also primitive and of the passive repeater type. In 1962 two more satellites were launched from Cape Canaveral; these were Telstar. They made history by being the first satellites to relay live television pictures across the Atlantic Ocean. These satellites orbited the earth once every 157 minutes in a low orbit. The first transatlantic telecast by Telstar was on 10 July 1962; it was a picture of the US flag fluttering in front of the sending station at Andover, Maine, USA. Thirteen days later telecasts showing life in the USA were relayed in Europe.

Neither Sputnik 1 nor Sputnik 2 fulfilled the prophecy of Arthur C. Clarke, in that they were not geostationary in orbit. They were experimental satellites for the purpose of probing space. Similarly, Echo 1 and Echo 2 launched by the USA were not of the geostationary type. Telestar was also not of a geostationary type but it did nevertheless represent a major advance in satellite technology because, unlike all other attempts at satellite design, Telstar was the first of the active-type satellites. It still needed an earth station to transmit signals to the satellite and receive signals from the satellite. Because the satellite appeared in the window of vision from earth for a limited time whilst passing through its orbit, the earth station had to be capable of tracking it until out of visual range.

The first experimental geosynchronous satellite was *Syncom*, launched from Cape Canaveral and used to relay the 1964 Olympic Games. Shortly after this event the CIS launched another satellite system called *Molniya* (lightning), but, like Telstar, it was not of the geosynchronous type. By this time both superpowers were gaining considerable experience in both rocket launching and satellite technology.

The first move towards legislation of satellite communications came from the USA in 1962 when it set up the Communications Satellite Corporation, which became known as *Comsat*. It was part private and part common carrier, and in part owned by the government. The policy laid down for Comsat was to set up satellite communication systems which could be used by governments, public corporations and privately owned companies.

By 1965 Comsat had expanded to include 15 member countries in addition to the principal member, the USA. That same year the organization was renamed *Intelsat* and launched its first satellite, *Early Bird*. This satellite, later to be known as *Intelsat I*, was launched from Cape Canaveral on 2 April 1965 and successfully parked at a height of 22 300 miles above the coast of Brazil in a geosynchronous orbit. This launch, unlike all other attempts, which were all experimental and not geosynchronous, marked the first major milestone towards the setting up of a worldwide network of geosynchronous satellites in space for the purpose of linking together the people of many nations.

History was made again in July of that yeár when the first live telecast via Intelsat satellite was made between the USA and the CIS. Early Bird represented a major advance on anything existing because it could carry 240 voice channels of speech. However, the transmission of a television signal requires a much greater bandwidth; a colour transmission needs at least 3 MHz. Therefore, in order to transmit television pictures in colour, all 240 voice channels had to be shut down as the bandwidth was needed for television.

Early Bird 1 provided a satellite link across the Atlantic Ocean, in a way repeating what Marconi had done with wireless telegraphy in 1901, but with vastly advanced technology. The next step in the concept of global communication by satellite was to link the USA across the Pacific Ocean to Hawaii and beyond; hence the use of monopole antennas on the early satellites. With such a wide beamwidth the transmitted energy is diffused over a large area. However, if the satellite needs to cover a much smaller area of the world then the beamwidth can be narrowed, with a corresponding greater gain on the system. Conversely, a smaller size of earth station could be employed.

This has given rise to the use of satellites for specific areas of the world, and these can be subdivided as follows:

1. Hemispherical coverage, covering a hemisphere, i.e. 20% of the earth.
2. Zonal coverage, usually defined as part of a continent.
3. Spot beam coverage; this concentrates the footprint of the satellite on a particular country of, say, Europe, or on a region of a large landmass such as North America.

Taking into account the global coverage systems such as that of Intelsat described earlier, each of these four satellite systems has a particular role to play in satellite communications. Global and hemispherical systems are used mainly by international users. Zonal coverage systems are favoured by the commercial satellite companies and find a use in satellite broadcasting in the USA, where a single SMATV receiving station can feed several thousand viewers by cable distribution.

Spot beam coverage is the system used in Europe for DBS.

Footprint

This is the primary service area covered by a satellite; the highest field intensity is normally at the centre of the footprint, reducing in intensity towards the outer edges.

The service area associated with any particular satellite's downlink is usually shown in terms of contour lines, each line representing a quality of service. This quality of service is sometimes expressed in transmitted equivalent radiated power in dBW, relative to an isotropic radiator. This is known as the effective isotropic radiated power (EIRP). Alternatively, the contour lines may show the received power flux density in dB (watts/m^2).

Transponder output power is usually the saturated power output level, whereas in practice transponders are normally operated with a 'back off', or power reduction to the linear portion of the travelling wave tube (TWT). The difference can be of significance, so users or leasers of satellites need to clear this with the owners of the satellite.

The shape and the area of a footprint can be varied at the design stage of the satellite, so as to provide coverage for a specific region of the world; it can be circular, kidney-shaped or oval, depending on the design of the downlink antenna on the satellite. Whilst it is possible to cover a theoretical maximum of 42% of the earth's surface with one satellite, the received signal strength would be very low, necessitating the employment of very powerful earth stations with very large receiving dishes to make up the shortfall in overall performance. Alternatively, the antenna on the satellite can be designed for gain at the expense of area coverage.

This is the design basis for all satellites providing a DTH DBS service to domestic viewers. Since it is neither economic nor practical for domestic viewers to use very large dishes, the gain must come from the satellite. The type of area coverage from a DBS satellite is termed a 'spot beam'. Because of the small area served by a DBS satellite, the effective power is directed to areas of high population density rather than the radiated power from the satellite being wasted over oceans and deserts.

The current generation of communication and DBS satellites are usually equipped with multi-beam (MB) systems which permit several beams to be transmitted simultaneously. A satellite may also have broad global beams and steerable highly directional spot beams. By the use of opposite polarization, a particular satellite frequency may be used for two, four or even six beams.

The centre of the spot beam where the maximum flux density occurs is known as the boresight.

Intelsat: an example of a satellite system

A satellite normally carries a number of transponders, the number varying from 3 to 19, or even 24, depending on the power in each transponder TWT, and the power available from the solar cells to drive the transponders to full power. Satellites are owned either by nations, consortia of nations, public companies owned by the shareholders, or simply by a private company. Examples of the first two categories are Aussat, the Australian satellite, and Intelsat, formed by a consortium of countries.

On 11 July 1967 Intelsat launched its second satellite; officially it was Intelsat II, but to many it was Early Bird 2. The word 'bird' has come to be increasingly associated with satellites. Intelsat II made it possible to transmit live television pictures across the Pacific, but to complete the vision of Arthur C. Clarke in 1945 there needed to be a third geosynchronous satellite, separated in space by 120° from each of the other two. In 1969 that third link was successfully put into place when the third satellite was placed in a geosynchronous orbit high above the Indian Ocean. Clarke's vision had finally come true.

But the research and development did not end there; technology goes out of date, and this is what happens to satellites just like any other piece of electronic hardware. Moreover, for a number of reasons, satellites have a finite lifetime, the main reason, assuming nothing else goes wrong, being the consumption of all their fuel.

Satellite coverage

The satellite communication system envisaged by Clarke was based upon the employment of three satellites in geosynchronous orbit, and spaced apart at intervals of 120°. Each satellite could in theory cover 42.4% of the earth's surface, and the coverage of any one satellite would overlap with that of each of the remaining two satellites. For coverage of the earth's surface from pole to pole, the satellite's transmitting and receiving antennas need to have a minimum beamwidth of 17.3°.

By the simplest definition a satellite is simply a repeater station having an uplink circuit from the ground station to the satellite, and a downlink from the satellite to the distant earth station, or to the customer's small earth terminal. Whilst in theory it is possible for one satellite to cover 40% of the earth's surface, in practice it is common to use satellites in tandem to effect a long-distance circuit. Thus, for example, a satellite transmission originating from New Zealand, intended for London, might use the Pacific Ocean Relay satellite of Intelsat, downlinked at Hong Kong and uplifted again, this time to the Indian Ocean Relay, where the signal is sent to London and downlinked by the British Telecom Earth Terminal.

Orbit geometry

Under the influence of gravity the orbit of a communications satellite can be defined as the trajectory described by its centre of gravity. When there are no perturbations, the satellite orbit describes a circle, ellipse, parabola, or hyperbola. For a closed orbit the trajectory is described in a periodic fashion which depends on the distance of the satellite from the centre of the earth, whether constant or variable (e.g. circular or elliptical orbit). The orbit is equatorial, polar or inclined, according to whether its plane contains the earth's equator, the earth's rotational axis, or is inclined towards the earth's equatorial plane.

However, a satellite with a constant speed in an inclined orbit requires continuous movement of the earth terminal antenna in order to follow the satellite in the sky, except in the unique case of the geostationary orbit. This orbit is a circular equatorial orbit in which the period of the sidereal revolution of the satellite is equal to the period of one sidereal rotation of the earth, and the direction of the satellite's revolution is in the same direction as the earth's rotation, that is, from west to east.

In summary, the geostationary orbit has the following main characteristics.

Period:

$$T_s = 23 \text{ hours } 56 \text{ minutes } 4.00954 \text{ seconds}$$

Radius:

$$R_s = 42\,165 \text{ km}$$

Satellite speed:

$$V_s = 3.0747 \text{ km/s}$$

Eccentricity:

$$e = 0$$

Inclination to earth's equator:

$$i = 0 \text{ degrees}$$

Altitude of satellite above equator:

$$h = 35\,787 \text{ km}$$

The geostationary satellite orbit has a simple geometrical relationship and simple properties compared with the non-geostationary satellite.

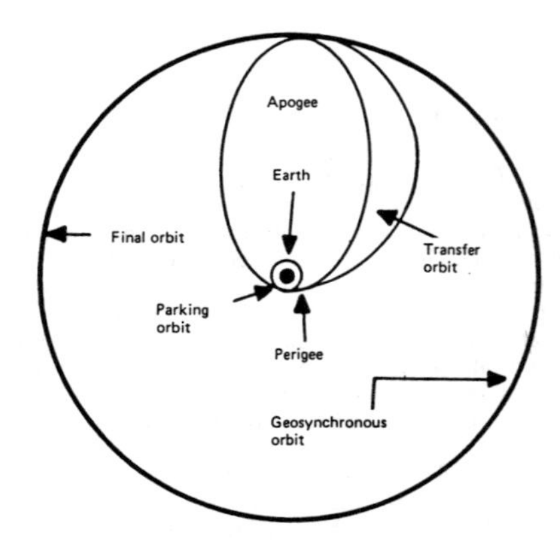

Figure 3.1 For low earth orbit, a rocket burn moves the craft from a parking orbit through transfer to the final orbit. To achieve a geosynchronous orbit, the satellite goes directly through the transfer orbit to the final altitude (Courtesy of Broadcast Engineering)

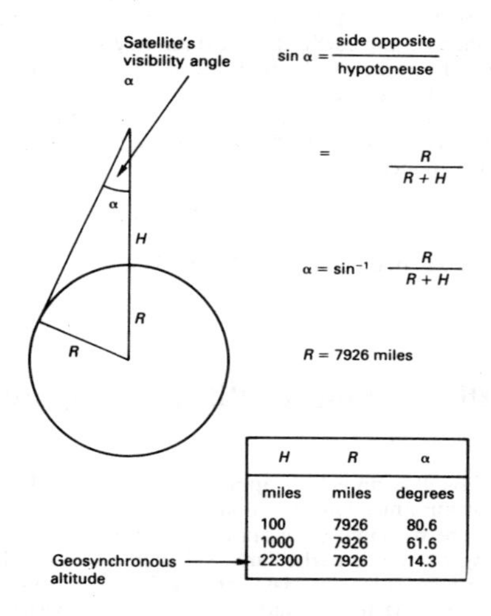

H	R	α
miles	miles	degrees
100	7926	80.6
1000	7926	61.6
22300	7926	14.3

Figure 3.2 Calculation of α, the angle of visibility (Courtesy of Daniel Leed)

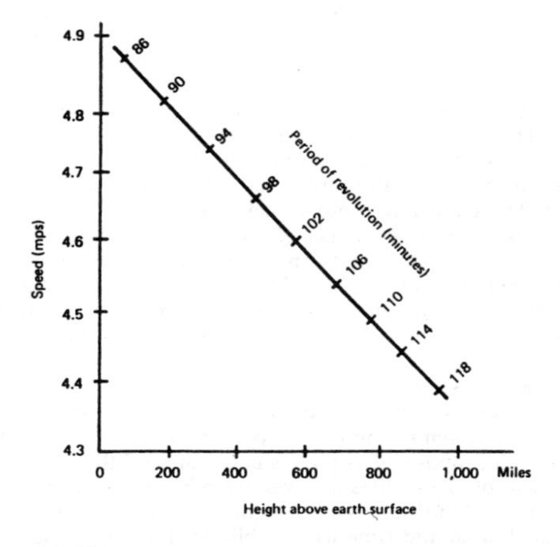

Figure 3.3 This nomograph shows satellite speed and period of revolution versus spacecraft altitude (Courtesy of Broadcast Engineering)

It is visible from all points on the surface of the earth except the polar caps (beyond 81.3°N and 81.3°S). When viewed from a given location on the earth, the geostationary satellite orbit would appear to be a curved line in the sky touching the local horizon at two different azimuths.

The radius R_s of a satellite in orbit is the sum of the equatorial radius of the earth, R, and the height of the satellite above earth, h, hence:

$$R_s = 6378.16 \text{ km} + 35\,787 \text{ km} = 42\,165 \text{ km}$$

The length of one degree of arc is 735 904 km.

Low earth-orbiting satellites (see Figure 3.3)

This type of satellite finds little application in present-day satellite telecommunications and none in DBS applications. However, there are some other applications for which the low earth-orbiting satellite is particularly well suited. One of these is in weather forecasting. In May 1991 the European satellite ERS-1 was finally blasted off into space, after many years of planning. The coming into service of ERS-1 marks the beginning of a new phase in collecting data about the earth.

Every 90 minutes ERS-1 makes one complete orbit at a height of 800 km. The satellite operates in what is called inclined orbit. This means that each pass over the earth sees a slightly different part of the earth's surface. Over several days a complete picture is built up of the weather pattern over the landmass and oceans, a synthesis quite impossible by any other means. From space ERS-1 measures sea waves, sea surface temperatures and ocean currents. Though calculation of ocean currents from the satellite is a complicated process, the principles are straightforward.

Winds blow parallel to the lines of equal pressure – the isobars. The closer the line concentration, the stronger are the winds. For oceans the equivalents of air pressures are the changes in sea temperature from the normal pattern. The currents flow parallel to these sea-level contours.

ERS-1 also maps the changes of the sea surface by timing the echo of the radar pulse to an accuracy of one-tenth of a thousand millionth of a second. ERS-1 is the first of a whole new generation of weather satellites which are going to bring some remarkable improvements in weather forecasting.

Mobile radio systems

Proposals have been made to use low earth orbit satellites for mobile radio systems. The US company Motorola has recently made such a proposal. The plan is based around the deployment of 77 satellites in seven equally-spaced orbits over the North and South poles at a height of 600 km. Imarsat, Canadian Telesat and American Mobile Sat have expressed a joint interest in backing this project with the name Iridium (the 77th element).

Each satellite would have a 3000-mile footprint addressed by 37 narrow beams. Satellite output transponder power would be lower than that required for geostationary satellites, thus reducing the size and weight of the satellite, with a corresponding saving in launch costs.

The Iridium project has the potential to be a bearer of international long-haul traffic, a profitable sector of worldwide telecommunications. Using fast inter-satellite switching at 60 GHz, the long echo associated with geostationary links (260 ms) would be avoided.

Each mobile could display time, latitude and longitude. Power on uplinks would be of the order of 7 W, voice coding will be the 8 kbits Motorola Codec and the proposal includes a data bearer at 2400 bit/s. The system proposed would use a frequency allocation subject to approval by WARC 1992, for a 10-MHz allocation between 2 GHz and 3 GHz.

Apart from the vast number of satellites deployed in the Iridium project, the scheme is complex in other areas of technology. It must be capable of accepting delivery of traffic from national networks and international networks for onward delivery to Iridium crosslinks and downlinks. The commercial viability of such a scheme has yet to be determined.

Amateur radio space communications

History tends to overlook the achievements of the amateur radio movement throughout the world which has been responsible for many major developments in radio communications. It was the American Radio Relay League, working in conjunction with its counterparts in France, Britain, Germany and the CIS (then USSR) that discovered the advantages of using the short waves, a part of the frequency spectrum which had been regarded as being of little value by commercial companies up to the late 1920s and early 1930s. The opening up of the frequency spectrum by the radio amateurs paved the way for HF long-range commercial communications.

Whilst it would be untrue to say that the amateur radio movement did the same thing with satellite communications, it is certainly true that the amateur radio movement was in the vanguard of satellite communications, using tiny satellites to achieve equally remarkable performance, enabling radio amateurs to communicate with each other through small transponders.

The data on amateur satellites in Table 3.1 is a concise potted history of low-power satellites and applications in amateur radio, scientific and educational studies.

Table 3.1 *Amateur satellites: a history (from the ARRL Handbook)*

OSCAR I the first of the phase I satellites, was launched on 12 December 1961. The 0.10 W transmitter on board discharged its batteries after only three weeks.

OSCAR II was launched on 2 June 1962. Identical to OSCAR I, Amateur Radio's second venture into space lasted 18 days.

OSCAR III, launched on 9 March 1965, was the first satellite for amateur communications. During its two-week life, more than 100 amateurs in 16 countries commucated through the linear transponder.

OSCAR IV was launched on 21 December 1965. This satellite had a mode-J linear transponder. A launch vehicle defect placed OSCAR IV into poor orbit, preventing widespread amateur use.

OSCAR 5, built by students at Melbourne University in Australia, transmitted telemetry on both 2 m and 10 m for more than a month.

OSCAR 6, the first of the phase II satellites, was launched on 15 October 1972. It carried a mode-A linear transponder and lasted nearly five years.

OSCAR 7 built by hams from many countries was launched on 15 November 1974. It carried mode-A and mode-B linear transponders and served the amateur community for six years.

OSCAR 8, again a co-operative international effort, was launched on 5 March 1978. Mode-A and mode-J transponders were carried on board. The spacecraft lasted six years.

Radio Sputniks 1 and 2, launched from the Soviet Union on 26 October 1978, each carried a sensitive mode-A transponder. Their useful lifetimes were only a few months.

OSCAR Phase III-A, the first of a new satellite series, was launched on 23 May 1980, but it failed to achieve orbit because of a failure in the launch vehicle.

OSCAR 9, built at the University of Surrey in England, was launched in October 1981. This is a scientific/educational low-orbit satellite containing many experiments and beacons, but no amateur transponders. UoSAT-OSCAR 9 re-entered the earth's atmosphere in 1989.

Radio Sputniks 3–9 were simultaneously launched aboard a single vehicle in December 1981. Several carried mode-A transponders and two carried a device nicknamed 'Robot' that could automatically handle a CW QSO.

Iskra 2 was launched manually from the Salyut 7 space station in May 1982 and sported a mode-K HF transponder. It was destroyed upon re-entering the atmosphere a few weeks after launch.

Iskra 3, launched in November 1982 from Salyut 7, was even shorter lived than its predecessor.

OSCAR 10, the second phase III satellite, was launched on 16 June 1983, aboard an ESA Ariane rocket, and was placed in an elliptical orbit. OSCAR 10 carries mode-B and mode-L transponders. OSCAR 10 is operational only part time as this is written. Listen to W1AW bulletins for operating schedules.

OSCAR 11, another scientific/educational low-orbit satellite like OSCAR 9, was built at the University of Surrey in England and launched on 1 March 1984. This spacecraft has also demonstrated the feasibility of store-and-forward packet digital communications and is fully operational as this is written.

OSCAR 12, built in Japan and launched by the Japanese in August 1986, is a low-orbit (phase II) satellite carrying mode-J and mode-JD (digital store-and-forward) transponders.

Radio Sputnik 10/11, launched in June 1987, is two separate 'satellites' on the same low-orbit spacecraft. The satellites carry mode-A, -K and -T transponders and CW QSO 'Robots.'

OSCAR 13, the third phase III satellite, was launched on 15 June 1988, aboard the first ESA Ariane 4 (an A44LP vehicle, to be specific) launch vehicle and placed in a near-Molniya, high-inclination orbit. OSCAR 13 carries transponders for modes B, J and L and also has experimental mode-S and RUDAK (packet) transponders.

OSCAR 14 and *OSCAR 15*, launched with four other Microsats (OSCARs 16–19) in January 1990, are the third and fourth amateur satellites built at the University of Surrey in England. Like UoSAT-OSCARs 9 and 11, they are designed for educational and scientific use.

OSCAR 16, also known as PACSAT, is a digital store-and-forward packet radio BBS. It had an experimental S-band beacon at 2401.100 MHz.

OSCAR 17, also known as DOVE (for Digital Orbiting Voice Encoder), was designed primarily for classroom use. Telemetry is downlinked using standard AX.25 packets using the Bell 202 standard. Experimental S-band beacon at 2401.200 MHz.

OSCAR 18, also known as WEBERSAT, contains scientific and educational experiments designed and built by faculty, students and volunteers at the Center for Aerospace Technology at Weber State University, Ogden, Utah.

OSCAR 19, also known as LUSAT, was sponsored by AMSAT Argentina and is nearly identical to AMSAT-OSCAR 16.

OSCAR 20, launched into low earth orbit in February 1990, is the second amateur satellite designed and built in Japan. It carries mode-J and mode-JD (digital store and forward) transponders.

4 Growth in satellite communication systems

The satellite operators

Such has been the remarkable growth rate in satellite systems that in a short time-span of just over 30 years, from the launching of the first Soviet-built Sputnik, we have witnessed the emergence of many different satellite communication systems, linking and connecting continent to continent, region to region, and country to country on a scale that no one could have dreamed of, even 20 years ago.

The satellite communication systems that have evolved can be classified as:

(a) major international systems owned by governments or consortia
(b) the nationally owned systems
(c) the privately owned carriers

In the first category there are three main international systems. These are:

1. Intelsat. Formed under the 1962 legislation by the USA.
2. Intersputnik. Formed by the CIS (then USSR).
3. Eutelsat. Formed in 1977 to link European countries.

In the second category there is a considerable number, increasing each year. These include:

1. Aussat. The Australian national system.
2. Arabsat. The Middle East system.
3. Brazilsat. The system developed by Brazil.
4. Insat. The Indian government-owned system.
5. Palapa. The Indonesian system covering the entire archipelago.
6. Mexico. The Morelos system.
7. Anik. The Canadian system.
8. France Telecom. Owned by France Telecom.

The third category includes the privately owned carriers; these are the commercially owned telecommunications giants like AT&T, and also the companies such as Hughes Corporation, manufacturers of satellites. There are to date four companies in the USA who have satellite systems; they are AT&T, Hughes Corporation, GTE and GE Americom.

All of these categories were designed to provide a variety of services, including telephony, data, and point-to-point television transmission.

Within the past decade there has been added a fourth category of satellites; those designed specifically for DTH broadcasting of television. These are styled DBS satellites. It is this category where the next growth will come from.

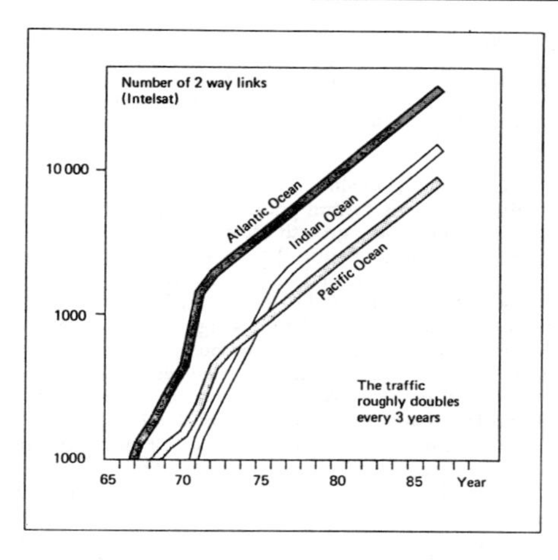

Figure 4.1 Growth of international satellite communications (Courtesy of Intelsat)

Figure 4.1 records the growth rate in satellite telecommunications traffic since the first Intelsat satellites went into service. It does not show the even more remarkable growth in DBS services which has taken place from 1988 onwards. Note the way in which the Indian Ocean traffic is approaching the level in Europe and America.

Business television by satellite

The link between business and satellite technology is a development of recent times. Like most innovative things it grew from small beginnings, fuelled by the advent of Ku-band satellites and small-aperture earth stations. Business television encompasses everything from one-way distribution of information to fully interactive activities like video conferences.

Table 4.1 *A comparison of earliest and recent generations of satellites*

	First launch of series			
	Early Bird	Intelsat VA	Intelsat VI	Intelsat VII
Year	1965	1980	1989	1991
Satellite mass in kg	138.5	900	1 870	1 425
Prime power in watts	40	1 200	2 200	3 900
No. of transponders	2	30	48	36
Total bandwidth (MHz)	50	2 160	3 303	2 300
Telephone channel capability	480	15 000	24 000	18 000
TV channel capacity		2	3	3
Suitability for small earth terminals	No	No	No	Yes

Table 4.2 *Type of satellite design*

Intelsat series V	Three-axis stabilized, cube construction 1.8 m
Intelsat series VI	Cylindrical with solar cells as part of body
Intelsat series VII	Three-axis stabilized, cube construction 1.8 m

The idea of using satellites to transmit visual and audio information from one central point to a large number of end-users has become a growth industry and like most business ideas it originated in the USA. In its simplest form it is non-interactive; all the end-user needs is a satellite receiving system with a medium-sized dish, 1.8 m. The sending of business information by satellite has significant advantages over other delivery methods such as post, telex and facsimile. The same package of information can be sent to hundreds of trading outlets, for example to supermarket stores, betting shops and travel agents. Apart from routine business traffic, the system can be used by the head office of an organization, enabling the chief executive to send instructions to the staff.

Non-interactive systems can take advantage of timeshifting. In the USA with its five time zones, a transmission from head office in New York can trigger the VCR machine at the remote end-user, record the instruction and screen at the appropriate time when the staff come on duty for the day.

Video conferencing is an example of a fully interactive system which needs an uplink at the satellite receiving system. Because of this, and the fact that it needs to have permanent studio facilities in cities, it is a business which is largely catered for by the major telecommunications authorities. Video conferencing can shrink the world; it is a lot easier to shift information than move businessmen around the world. It leads to time saving and increased efficiency.

The advent of the Ku band, which enables the employment of medium-sized earth stations, paved the way for video conferencing. Conferences are not limited to connecting cities in a country; today, conferences can be conducted from almost any major city in Europe to other cities in the USA, Canada, Japan, Singapore, Hong Kong and Australia. Neither is video conferencing limited to two parties; as many as 12 studios in different parts of the world can be linked on a fully interactive basis. The recent political changes in Europe have contributed to a potentially large market for Western manufacturing companies.

British Telecom, France Telecom and Telecom Finland are only three of Europe's telecommunications authorities who are seeking to capture a share of the satellite business communications market. British Telecom entered the field as a commercial provider of television and video distribution in 1984. Its first role was to distribute television programmes by satellite to head-ends of television cable companies. Since then the original single satellite dish at its London teleport has been supplemented by a number of dishes ranging in size to 13 m. Satellite services include television programme distribution, digital services and video conferencing. The main satellites accessed by BT Visual Broadcast Services are Astra, Eutelsat, Intelsat and PanAmSat.

Satellite market profile

The global launch market for all types of payloads over the past five or six years was 100–120 payloads for 85–95 launches according to the annual market update carried out by Arianespace. The commercial market represents only about 15% of the total market, since all CIS satellites and all US military satellites are launched from national launching resources. In the case of communication satellites it is a different story, the launch contract being awarded to the most successful bidder.

For the world's space communications and those satellites launched solely for the purpose of DBS, the world total of satellites in geostationary orbit comes to about 150; these are managed by 30–35 operators with a total of about 1850 transponders.

The distribution of television programmes to cable network operators and to DTH viewers is fast becoming the major growth market for satellites. Arianespace predicts a market surge in satellite launches for 1994–1995 when the US satellite network will have to be renewed, and when the last phase of the European network will occur. From 1996 to 1998 it sees its market as a steady renewal of international systems.

The major networks

Intelsat

The International Telecommunications Organization (Intelsat) was born in 1964. The forming of this international organization supplied ample proof of the importance of satellites in space. It was set up to administer and regulate global satellite coverage for member nations. One year later it launched Early Bird, the first of a long line of geostationary satellites. Today there are 117 signatories to the international agreement. The first satellite had a limited capacity of 240 voice channels and one television channel; today Intelsat has at least 16 satellites in operational use and since the first launch in 1965 it has launched 33 satellites constituting five generations of satellite technology, the latest having a capacity of over 30 000 voice circuits and three television channels.

The fifth generation of satellites which Intelsat began launching from 1980 were physically much bigger than Early Bird and their high channel capacity meant that they were the mainstay of the fleet until 1989 when the sixth generation came into service. Intelsat VI was one of the biggest satellites to see service. It had a capacity of 270 000 all-digital telephone channels and 48 000 analogue telephone channels. Conversely it can accommodate more television channels and fewer telephone channels.

Intelsat comes into its own when major international events are being staged, as well as providing day-to-day telephone, telegraph, data and television to 117 different countries of the world.

Intelsat operates satellites in a geostationary orbit 36 000 km above each of the three main regions of the world, i.e. above the Atlantic Ocean, the Pacific Ocean and the Indian Ocean. These are designated AOR, POR and IOR respectively. From these locations the satellites are ideally situated to carry telecommunications traffic to all parts of the world, conveniently situated as large landmasses on either side of the oceans.

According to analytical studies this traffic is increasing at a rate of 15–20% per annum. To meet this demand Intelsat invited bids for a seventh generation of satellites. Five Mark VII satellites have been ordered and the first of this new series, Intelsat K, was launched into geotransfer orbit from the first Atlas IIA launch vehicle in July 1992.

Intelsat, in addition to providing satellites for a global communications system, also operates eight telemetry, tracking and control command (TTC) stations throughout the world. These in turn are controlled by Intelsat Operations and Spacecraft Control Center, Washington DC.

Satellite design The Intelsat VIIs will be of a different design to the VI. The VII spacecraft is a three-axis stabilized type, comprising a cube of 1.8 m on all sides, and attached to this satellite bus are the two solar panels. This is not the same as the VI, which is spin-stabilized and cylindrical in shape. The design and construction of satellites is an ongoing process. Each particular design approach has its particular merits. All have a common objective; to increase operational life, to increase flexibility in operation, to increase capacity of traffic handling, and finally to permit flexibility in terms of being able to launch from any one of a number of vehicles.

Intelsat K Intelsat K is a special version of the Series VII of Intelsat satellites designed for television traffic. Intelsat K is an impressive bird which puts out 3155 W of transponder power spread over its all Ku transponders. Each delivers up to 60 W and there are additional transponders to allow for redundancy. Each of the transponders feeds into dual antennas of the linear vertical and linear horizontal mode. The suffix K is derived from the fact that this is the first of its kind intended for broadcasting in the Ku band.

Uplink frequencies are in the 14–14.5 GHz band, and downlink frequencies are in the 11.45–11.7 GHz band, with 11.7– 11.95 GHz for North and South America, and 12.5–12.75 for Europe. Intelsat K allows for simultaneous uplinking for either point to point, point to multipoint, or multipoint to multipoint transmissions. The broadcast capability permits the user to uplink separately or even simultaneously from both Europe and the USA.

Intelsat K's transponders deliver footprints between 47 and 50 dBW which puts it on equal terms with most of the DBS satellites in Europe. The transponders give a picture quality of the highest standard in linearity in any of the current terrestrial television formats, NTSC, SECAM and PAL. In addition, transmissions in B-MAC and D-MAC can also be accepted.

Intelsat K has set a benchmark in satellite technology in that it is the first satellite from an international common carrier that is able to provide a service to broadcasting stations anywhere in the eastern portion of the USA, parts of South America, and the whole of Europe.

Today Intelsat has the largest satellite system in the world, and the only one with a truly global capability. In 1992 it had 122 nations as shareholders and it is growing each year (Table 4.3). Revenue earned in 1991 was $US 563.4 million.

In interpreting these figures it should be noted that some country's holding may be low because they prefer to invest in their own satellite systems, e.g. France.

Intelsat has plans for a series VIII: a contract for two has been signed with GE Astrospace (Table 4.4).

Table 4.3 *Major shareholders and percentages*

Country	*Holding*
USA	21.86
UK	12.05
Japan	4.5
France	4.2
Germany	4.19
Australia	2.7
Italy	2.4
Singapore	2.0

Scope of Intelsat

Intelsat currently provides global telecommunications via 18 satellites; system reliability is claimed to be 99.999%. Intelsat K in the series VII is Ku-band.

Arabsat

The past decade has witnessed extraordinary growth in satellite communications, and the greatest growth rate has taken place in a region of the world which no more than 50 years ago regarded radio as the voice of the devil. This area of the world is that which stretches from the Arabian Gulf to the western end of North Africa; the Arab world. In many spheres of technology the Arab states are in the vanguard. One of these is in telecommunications and satellite systems.

In 1976 the League of Arab States (LAS) formed the Arab Satellite Communications Organization (Arabsat). Its purpose was to provide a number of telecommunications services to 21 Arab countries, having a total population of around 200 million people distributed over a region of the world extending to 14 million square kilometres. The core of this telecommunications network is its satellite system, Arabsat. From the outset the new leaders of the Arab world were quick to perceive the strategic advantages of such a system. To a region of the world characterized by small cities and villages cut off from twentieth-century civilization the coming of satellite communications made it possible not only to provide a modern, efficient network for telephone, telegraph and data, but also to link up the more remote communities with television, offering educational opportunities, entertainment, and news and views from the major cities of the Arab world.

The Arabsat network was formed initially as a system to complement that of Intelsat but it quickly developed, aided by the wealth of the Arab oil-producing states, into one of the most modern and best equipped telecommunications networks of the world.

Table 4.4 *Current and future Intelsat spacecraft*

Intelsat Designation	*Intelsat V*	*Intelsat V-A*	*Intelsat VI*	*Intelsat K*	*Intelsat VII*	*Intelsat VII-A*
Year of first launch	1980	1985	1989	1992	1993	1995
Prime contractor	Ford Aerospace	Ford Aerospace	Hughes	GE Astro Space	SS/Loral*	SS/Loral
Launch vehicles	Atlas Centaur Ariane 1, 2	Atlas Centaur Ariane 1, 2	Ariane 4, Titan	Atlas IIA	Ariane 4, Atlas IIAS	Ariane 44L
Lifetime (years)	7	7	13	10	10–15	10–15
Capacity	12 000 circuits and 2 TV	15 000 and 2 TV	24 000 and 3 TV (up to 120 000 with digital circuit multiplication equipment, DCME)	16 54 MHz Ku-band transponders; can be configured to provide up to 32 high quality TV channels	18 000 and 3 TV (up to 90 000 with DCME)	22 500 and 3 TV (up to 112 500 with DCME)

*Formerly Ford Aerospace

Arabsat launched two satellites in 1985, an event of some magnitude, for in the timescale of nine years since the Arab leaders first met to form Arabsat, the Arab world had moved forward 100 years in time to the twentieth century. The first satellite was launched in February 1985 and the second in June 1985 in geostationary orbits of 19°E, and 26°W respectively with characteristics as shown in Table 4.5.

Table 4.5

Type	Three-axis stabilized
Frequency bands	Rx 5925–6425 MHz Tx 3700–4200 MHz } C-band Rx 2500–2690 MHz
Transponders	25 in C band 1 in S band
Transponder bandwidth	33 MHz
EIRP	31 dBW (nominal 34–26) C band 41 dBW at beam edge of S band
Total capacity	8000 telephone channels 7 television channels

Note: Rx = receiver, Tx = transmitter
Source: B. Ackroyd, *World Satellite Communications and Earth Station Design*

The Arabsat satellites were manufactured by Aerospatiale. The construction is a cube configuration with two projecting solar panels giving a total body length of 20.7 m and a payload in orbit of 1000 kg. The satellite has a design life of seven years.

The second important element in the Arabsat system is the ground control. Arabsat has its main tracking and control station at Riyadh. This performs two vital functions or phases. These are the launch phase and the operating phase. During the critical period of the launch, Riyadh monitors and controls the orbital position and the commissioning of the satellite. The next phase is that of maintaining the satellite in its correct orbit. This is where telemetry data from the satellite are received and processed, in order that repositioning instructions can be given to the satellite should it deviate.

Because of the importance of ground control, the Riyadh station (the primary site) is backed by another station located at a site near Tunis. Both stations are fully operational, and each has fully redundant equipment. The telemetry receive frequency is either 3703.7 MHz or 4199.9 MHz, with 5925.5 MHz as the command frequency uplink to the satellite.

The third element in the Arabsat system is the network of ground stations. Each of the countries making up Arabsat has at least one earth station. The main stations operate in the C band and are capable of transmitting and receiving. Basically there are four types of earth stations:

1. Main earth stations operating between large metropolitan areas. These handle a large number of telephony channels and may be regarded as the backbone of Arabsat. These earth stations can originate and receive television programmes.

2. Medium-power earth stations. These have a lower channel capacity and can receive television programmes but cannot transmit television.
3. Earth stations for emergency communications. These are fully transportable, complete with power supply systems on flat-bed trucks. Some have a television facility.
4. Small earth stations for receiving community television and re-broadcasting by VHF.

Because all Arab nations are linked by culture, history, traditions, religion and a common tongue that extends from Morocco to Oman, a television programme coming from the studios in one country can be enjoyed with equal pleasure in all Arab and Arab-linked countries. For these reasons satellite broadcasting of television is likely to assume a greater importance in the future.

In 1992 Arabsat awarded a contract to Hughes Aircraft Company, following an international tender, for two second-generation satellites. These will be the highly successful Hughes HS 601s. Each 601 will carry 20 S-band and 12 Ku-band transponders for the purpose of providing enhanced coverage throughout the Arab world. DBS reception will be possible using 68–80 cm sized dishes. The launch vehicle will be the Ariane 4.

Eutelsat

Eutelsat was a consequence of the forming of the ESA, and was conceived with the idea of linking together by satellites the telecommunication networks of the member countries of the European Space Agency (ESA). The planning of Eutelsat got under way in 1977, with the objective of producing a satellite system that would cover an area extending from Scandinavia, to Yugoslavia in the east, to North Africa in the south, and as far west as Ireland and Iceland.

Eutelsat was conceived on similar lines to Intelsat but with the main difference that Eutelsat would use the new Ku band so as to avoid interference with Intelsat. The use of the Ku band offered prospects of greater capacity together with a number of other advantages such as smaller earth stations in place of the massive structures used at some Intelsat ground stations.

After several years of planning, Eutelsat launched its first satellite in 1983, an event which signified the beginning of a new era in the combination of space age technology with a revolution in telecommunications and television broadcasting. Since that first launch, Eutelsat has set up a satellite communications network that takes in nine satellites to date, with more planned for the future (Table 4.6).

The Eutelsat network of satellites is designed to link all members of Eutelsat (Table 4.7). These countries are also members of the Conference Européenne Des Administration des Postes et des Telecommunications (CEPT). Each country is required to provide an earth station as the point of entry. Eutelsat carries telephony, telex and data streams between members. It also carries out television broadcasting. Eutelsat has an ongoing special relationship with the European Broadcasting Union, providing transponders for television link-ups between member countries. The past ten years has witnessed a rapid growth in the use of satellites for SMATV.

Table 4.6 *Eutelsat satellite network*

	Orbit	Launch
Eutelsat I F1	13°E	1983
Eutelsat I F2	7°E	1984
Eutelsat I F3	10°E	1985
Eutelsat I F4	7°E	1987
Eutelsat I F5	7°E	1988
Eutelsat II F1	7°E	1989
Eutelsat II F2	7°E	1990
Eutelsat II F3	7°E	1991
Eutelsat II F4	7°E	1992

Table 4.7 *Member countries of Eutelsat*

Austria	Iceland	Portugal
Belgium	Ireland	San Marino
Cyprus	Italy	Spain
Denmark	Liechtenstein	Sweden
Finland	Luxembourg	Switzerland
France	Malta	Turkey
UK	Monaco	Vatican City
Greece	Netherlands	Germany
	Norway	Yugoslavia

Note: As a consequence of the break up of the Soviet Union and the Eastern Bloc, the number of countries joining Eutelsat has increased to 38.

With the introduction of DBS it was logical that Eutelsat would play a major role in Europe.

Eutelsat was conceived long before DBS came into existence, but will become a DBS broadcaster by the year 1996. Member countries pledged their support for Eutelsat to operate its own DBS satellites. The project is named Europesat and the agreement was signed in December 1990. Bids were invited in July 1991 and a contract award is scheduled.

Europesat will operate a three-satellite system plus one spare and all will operate from the same position at 19°W. The system is intended to offer a series of steerable spot beams for regional and linguistic markets within Europe. Some 39 transponders will come into operation initially, with more to follow later.

The Eutelsat Consortium is one of the few satellite operators in Europe which intends to support the principle of using high-power transponders so as to provide the viewer with a better quality of picture.

Aussat/Optus

The Australian National Satellite Systems Operating Company was formed in 1986. Aussat was the result of ten years of system planning by the National Satellite Task Force set up for that purpose. In its initial concept the satellite system covered not only Australia but also Papua New Guinea and New Zealand. In the final scheme New Zealand opted out. In 1993 the name Aussat was replaced with Optus. References to these are therefore synonymous.

For Australia, with its massive landmass characterized by cities and towns separated from one another by vast expanses of hostile desert, and remote areas of a few hundred homesteads, the coming of satellite communications provided the ideal solution to its communication difficulties. Australia has this much in common with the Middle East countries such as Saudi Arabia. For the first time it brought television within reach of many thousands of its citizens who previously had to rely upon shortwave radio for long-distance reception.

Australia has eight main cities: Adelaide, Brisbane, Canberra, Darwin, Hobart, Melbourne, Perth and Sydney. All of these cities have their own terrestrial television stations but because of the great physical separation between some of the cities it meant that viewers had no choice of television programmes; satellite television has corrected this. Adelaide, Darwin, Perth and Sydney each have two earth stations, and the rest have one. The Aussat system is designed to provide a number of services in addition to television. The main ones are long-range telephone circuits, and special service broadcasting (SSB) intended to replace or act as an adjunct to the shortwave system which is still in service.

The Aussat system uses three satellites, all of the geostationary type with identical performance parameters.

Australia's second national satellite system In July 1988, AUSSAT Pty Ltd, Australia's national satellite communications company, became the first customer to purchase the Hughes Aircraft Company HS 601 body-stabilized satellite when it ordered two of the high power spacecraft for its next generation Aussat B system.

Built by the Hughes Space and Communications Group in El Segundo, California, the HS 601 is considerably more powerful and versatile than previous Hughes satellites. Aussat B (Figure 4.2) is three times more powerful and will last twice as long as Aussat A, Australia's first national communications satellite system, which was also built by Hughes. The Aussat B satellites will enhance existing satellite communications services throughout Australia, including direct television broadcast to homesteads and remote communities, voice communications to urban and rural areas, digital data transmissions, high quality television relays between major cities, and centralized air traffic control services.

In addition, Aussat B will introduce the first domestic mobile satellite communications network to Australia. The satellite is equipped with a 150 watt L-band transponder that will permit mobile communications through small antennas mounted on cars, trucks, and airplanes. This mobile ability will extend throughout the nation, permitting travellers to maintain communications from moving vehicles.

The satellites will also use high powered spot beams covering the major cities to provide such specialized services as high performance data links, video conferencing, and a range of other dedicated services, including direct broadcast for pay TV.

Aussat B is a body-stabilized, three-axis design that consists of a cube-shaped central body, 2.29 m each side, with a pair of three panel solar array wings. Each wing extends 9.14 m north and south from the body for an overall deployed length of 20.57 m. A 30-element L-band antenna array covers the earth-facing surface of the spacecraft. One oval reflector deploys from the east side of the spacecraft body; and two smaller oval reflectors on the west side are attached to an A-frame structure, similar to that on Aussat A. The three-reflector antenna system provides seven transmit and three receive beams in vertical and horizontal polarization. The satellite weight at beginning of life in orbit will be 1582 kg.

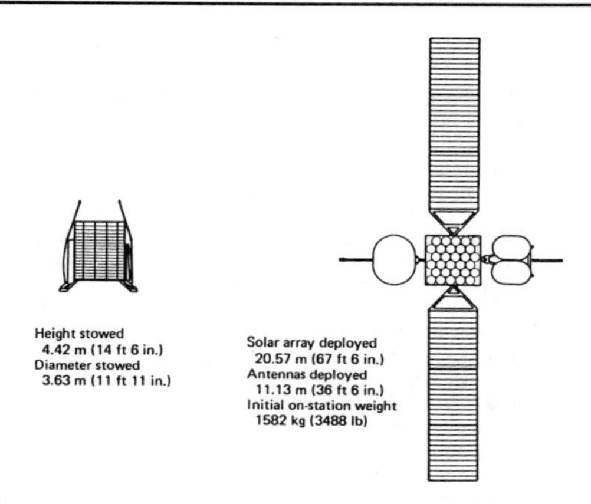

Figure 4.2 Aussat B (Courtesy of Hughes)

The Ku-band communications payload consists of fifteen 50 watt linearized transponders, each with a bandwidth of 54 MHz. The transmit coverage includes two national beams to all of Australia and the offshore region; spot beams to the western, central, north-east, and south-east regions of the Australian continent; a national beam to New Zealand; and a trans Tasman beam. It will be possible to switch eight of the transponders on each satellite to provide domestic service to New Zealand. The effective isotropic radiated power (EIRP) varies from 44 dBW to 51 dBW, depending on the beam.

In addition to the Ku- and L-band transponders, Aussat B will carry two experimental payloads, a Ka-band beacon and a laser retroreflector. Both experimental payloads are located in the L-band antenna, where they have the required visibility of Australia. The Ka-band beacon will transmit a 28 GHz signal on both horizontal and vertical polarizations for propagation experiments. The laser retroreflector will permit precise location of the spacecraft so that signals sent through Aussat B can be used to set timing standards throughout Australia.

The electrical power subsystem uses two sun tracking solar arrays to generate 3200 watts of electricity. The three panel solar array wings are covered with large area K4-3/4 silicon solar cells. Each panel is 2.54 m by 2.16 m. A 28 cell nickel-hydrogen battery provides full power to the spacecraft during eclipse operations when the satellite passes through the earth's shadow.

The satellite's integral propulsion system carries 1658 kg of monomethyl hydrazine and nitrogen tetroxide bipropellant in four spherical titanium tanks, providing an in-orbit life of 15 years. A single 490 newton (110 pound force) thruster is used for perigee augmentation and apogee burns, and thirteen 22 newton (5 pound force) thrusters are used for attitude control and stationkeeping manoeuvres. Highly accurate antenna pointing control is provided by independent beacon tracking on each of the Ku-band reflector systems. The attitude control system uses an innovative combination of double gimballed momentum wheels and solar tracking, which minimizes the need for thruster usage during normal on-station operations.

The first Aussat B satellite was launched in August 1992 from a Chinese Long March rocket. The next satellite, B2, is scheduled for launch in 1994. The Long March will place the satellite in a low earth orbit and impart a slow spin for stability during orbital transfer manoeuvres. The satellite's onboard Thiokol Star 63F perigee kick motor (PKM) will be fired and will boost the spacecraft into a geosynchronous transfer orbit. After its solid fuel is depleted, the PKM will be jettisoned. The liquid bipropellant apogee motor using the 490 newton thruster is fired by ground command on three separate apogees, with each firing raising the perigee of the transfer orbit until the orbit is circularized at geosynchronous altitude 36 000 km above the equator. As the satellite drifts toward its assigned orbital position, it is despun; the reflectors are deployed; and the solar panels are extended. The spacecraft is oriented and the momentum wheel is activated. Aussat B1 is positioned at 160°E longitude, and Aussat B2 will be positioned at 156°E longitude.

The Australian satellite Optus B2 suffered some damage after its launch and had to be declared a total loss. The Chinese Launch Agency maintained that their launch put the spacecraft exactly where planned, in a 213 × 1032 km orbit, and that the cause of the loss was with the spacecraft. The cost of a replacement spacecraft and its launch will be covered by the underwriters. It is expected the replacement 601 will be launched in 1994 from another Chinese rocket, LM-2E.

Brazilsat

Brazil is another country whose large landmass makes it ideal for a solution to its communication problem to be based on satellites. Brazil initially bought time on the Intelsat system but now has its own satellites. The system uses C band, and has a capacity of 12 000 telephone channels or 24 television channels. Brazilsat is based on C-band satellites and uses the system for the following services:

(a) transmission of television between transmitter stations
(b) distribution of radio programmes
(c) telephone circuits
(d) direct broadcasting of television

Brazilsat has placed an order for two second-generation satellites. These will be wide bodied versions of the HS-376, more powerful than the early 376s. Due for delivery and launching in 1994 they will be the result of a joint engineering agreement between the US and Brazil.

Insat

This is the Indian government-owned national satellite network. The system came into service in 1983, though initially the Indian government leased time on the Intelsat network before acquiring its own satellites, which are C-band technology. The service is a fully integrated one that satisfies the needs of three types of service: telecommunications, television and meteorology.

Palapa

Indonesia is a fragmented archipelago of islands that stretches over a distance of 1500 miles in the south Pacific. With much to gain in infrastructure by investing in the most modern form of telecommunications the Indonesian government was one of the first countries in the world to invest in a satellite communications network.

The system, conceived in the early 1970s, was actually inaugurated in 1976 with the launching of its first satellite designated Palapa A01. In the subsequent years the system was expanded with the launching of a further three satellites, all manufactured by Hughes. Conforming to the state of the art of that time, all are operating in C band, with TWTA output powers up to a maximum of 10 watts. However the last two satellites have a much greater channel capacity as shown in Table 4.8. Satellites A01 and A02 were launched respectively in 1976 and 1977 at orbital positions of 83°E and 77°E, whilst B3 and B4 were launched respectively in 1983 and 1986 in orbital positions of 113°E and 118°E. The satellite B3 was a replacement for the earlier satellite B2, launched in 1984, which failed to reach orbit.

Table 4.8 *Palapa satellite network characteristics*

Transponders	*AO1*	*AO2*	*B1*	*B3*
Number	12	12	24	24
Bandwidth MHz	36	36	36	36
EIRP dBW	33	33	3	36
TWTA power W	5	5	10	10
Channel capacity	6000	6000	24 000	24 000
Speech (or)	6000	6000	24 000	24 000
Television	12	12	24	24

Source: B. Ackroyd, *World Satellite Communications and Earth Station Design*

Indicative of the expanding role that Indonesia intends to play in the future in satellite communications and broadcasting, is that it has already hosted an international convention in space communications during 1991. Eventually Indonesian satellites will offer services to Thailand and Korea and to other countries on the Asia-Pacific rim.

Palapa B4 was successfully launched in May 1992 from a McDonnell-Douglas Delta 2 rocket, version 7925. Palapa B4 will provide telephone, data and television signals to most of the 13 000 islands forming Indonesia. Palapa B4 is the last of the B series, a Hughes HS 376 with 24 C-band transponders capable of carrying 1000 duplex voice circuits or one TV channel. The B series are four times more powerful than the Palapa A0 series, HS 333s also built by Hughes Space and Communications Group. Palapa B4 orbits at 97.5°E.

Indonesia awarded a further contract in 1993 to Hughes, for the Palapa C network. Hughes will supply two HS 601 satellites with an option on a third for 1999. Thus the US company will have supplied Indonesia with three generations of spacecraft; the first being the HS 333, the second being the now famous HS 376, of which more have been built than of any other satellite in the world, and now, the HS 601, for which Hughes, to date, has 32 firm orders.

These latest satellites have considerably more power than the HS 376. With their 34 transponders, the two HS 601 will provide enhanced satellite service to all parts of Indonesia, southeast Asia and parts of China and Australia. It is not by coincidence that Indonesia was the second nation on earth to invest in a national satellite communications network: with its 13 000 islands, there is no other form of communications that could provide an effective and efficient telecommunications and television relay infrastructure.

Morelos network

Mexico moved towards having its own satellite system a few years ago. Due to its close connections with the USA, the country quickly moved to satellites with the purchase of two dual band satellites (C band and K band) from Hughes. These are Morelos 1 and 2 which provide a good infrastructure for nationwide television. Now Mexico is expanding the service with the purchase of another two satellites from Hughes, to form a network called Solidaridad. Mexico, a country with 86 million people, has one of the greatest concentrations of satellite dishes per capita of any country in the world.

Further indication of the way South American countries have embraced satellite communications is supplied by the contract placed by Mexico for a further two satellites of the HS 601 type.

Canadian satellite systems

Canada is another large landmass of mountains and large plains with the major cities being separated by long distances in many cases. Thus, as in the case of Australia, the coming of satellite communications provided an ideal solution. However, though Australia was a relative latecomer to satellite technology, this was certainly not the case with Canada. Canada has always been in the vanguard of satellite technology.

In 1972 it launched its first satellite, Anik 1, followed one year later by Anik 2. Both these satellites are time-expired and have been replaced with the second generation of satellites, Anik B, Anik C and Anik D. Anik B uses C band and Ku band, whilst Anik C is Ku band and Anik D is all C band. This combination of three different types of satellites enables the system to have a high degree of flexibility. The Anik C, which is all Ku band, has 15 transponders, each with a bandwidth of 54 MHz, and which achieve an EIRP of 49 dBW at the centre of the footprint.

Satellite systems in the CIS

From a position of being the first nation in the world to launch a satellite into space, the CIS (then USSR) gave priority to ensuring that it held onto its prestige in space sciences. Its first move was to form a satellite system which rivalled that of Intelsat; this was called Intersputnik. Like Intelsat, it was based on C-band technology because at that time in the 1960s the Ku band had not been explored. The system was designed to link together not only the landmass of the CIS, but also the countries in the Warsaw Pact, and the countries in the Middle East, Far East and Central America which had political ties with the CIS (Table 4.9). Later on, Intersputnik was expanded to open its services to any country.

Table 4.9 *Former members of Intersputnik*

Europe	*Middle East*	*Asia*	*Central America*
Bulgaria	Libya	Laos	Cuba
Czechoslovakia	Syria	Mongolia	Nicaragua
Germany	Yemen PDR	Vietnam	
Hungary	Afghanistan	North Korea	
Poland	Algeria	CIS	
Romania	Iraq		

The CIS uses a variety of satellites, identified as Stationar, Gorizont, Raduga and Ekran. These are of different generations and different operating frequencies such that the total CIS satellite system has a capability which includes geosynchronous satellites, non-geosynchronous satellites and coverage of C band, Ku band, L band and UHF, with Ka band to be added later. Stationar satellites use C band only, and transmit telephony, data and television. The Gorizont satellites use C band but with higher powered TWTAs of 40 W. The Raduga generation of satellites came into service in 1975, and were similar in design to the Gorizont type. Ekran satellites were different in the sense that they were intended for use in DBS. The downlink frequency was in the UHF-TV band 706–726 MHz.

The end of the Cold War revealed that Russia is ahead of the West in many areas of space technology. It is now known that the Russians built and launched over 1500 spacecraft of one type or another, some of which are of an advanced design (Table 4.10).

Western European countries

France is the hub of the European Space Agency, and also has its two satellite systems. The first of these was the Telecom satellite system, owned and operated by France Telecom. This is a national network completely independent of any international system. The purpose of this network is to provide a number of different services: telephony and digital data to the Caribbean, Canada and French Guiana. Additionally, the satellites carry television channels. Telecom F1 uses C band and also Ku band. A second satellite, F2, was launched a year after F1, in 1985. F2 is all Ku-band transponders. It has a spot beam centred on France, with an EIRP of 50 dBW at the centre of the footprint. The beam edges stretch to Scandinavia and North Africa with a reduced EIRP to about 41 dBW.

A second satellite system operated by France is that belonging to the state-owned TeleDiffusion de France (TDF). In 1990 it launched its first satellite, a high-power transponder intended for DBS. This was TDF 1. One year later in 1991 TDF launched its second DBS satellite, TDF 2.

These satellites are the first satellites to carry multiplexed analogue component transmission by D2-MAC in Europe.

France is the technological powerhouse of Europe, its giant company Thomson SA working in close cooperation with France Telecom and TeleDiffusion de France. France Telecom also has three research centres; the Combined Study for TV Broadcasting (CCETT), the National Telecommunications Study Centre (CNET), and the TeleDiffusion de France Research Centre (CERIM) dedicated to HDTV transmission.

Table 4.10 *Current CIS civilian communications satellite systems*

	Constellation	*First launch*	*Mass (kg)*	*Power (kW)*	*Lifetime (years)*	*Payload Transponders*	*Power (W)*	*Frequency Transmit (GHz)*	*Bands Receive (GHz)*	*Mission*
Molnya 1	High-inclination LEO (one plane) 8 satellites	1965	1800	0.04	3	1	40	0.8	n/a	Government communications
Molnya 3	High-inclination LEO (one plane) 4 satellites	1974	1800	1.0	3	2–3	40–80	4	6	FSS, television distribution
Raduga	Geostationary 8 satellites	1975	1965	n/a	3	n/a	n/a	4 7 11 1.5	6 8 14 1.6	Government communications
Ekran	Geostationary 1 satellite	1976	2000	0.2	3	1	200	0.7	n/a	DBS
Gorizont	Geostationary 8 satellites	1978	2120	1.3	3	6 1 1	34–40 n/a n/a	3.6–3.9 11.5 1.5	5.9–6.3 14.3 1.6	FSS, DBS MSS
Lutsch	Geostationary 3 satellites	1985	2400	1.8	5	3	240	11/13	14/15	Data relay, some DBS

Source: Via Satellite

Growth of the commercials

It was inevitable that the commercial operators would seize upon the opportunities that satellite communications offered, and by the mid-1970s commercial companies moved into the picture. But some of these were no ordinary companies. they were the American giants, companies such as AT&T, RCA and Western Union, the very same companies who had pioneered the cable telegraph systems of the (first) wire age. With the coming of the giant wireless telegraph transmitters, these same companies diversified into an alternative means of communication, and when the short waves were discovered these companies quickly exploited another new medium of communications. It was therefore inevitable and fitting that these very same companies should seek to be in the vanguard of satellite communications.

The logic of these giant companies having their own satellite systems to complement their terrestrial-based communication networks is self-evident. The American companies who operate satellite communication systems are shown in Table 4.11.

Domestic satellites, usually abbreviated *domsats*, perform a wide range of essential services and they are collectively part of America's strategic satellite communications network. Their area of coverage is by no means restricted to the Conus region (continental America): they reach out as far as the Caribbean and Hawaii. Domsats carry traffic ranging from television relay services for America's public service broadcasters to the US Armed Forces radio and TV networks. They act as carriers for certain US government agencies and also provide the essential services to America's many hotel chains e.g. room reservations and bookings, particularly to Hawaii and Florida where hotel density is at its highest.

Domsats

AT&T, the largest public service telecommunications network in the world, is also historically the first satellite operator in the world. The first US satellites launched in early 1960 were AT&T owned, and later AT&T formed its satellite network called Telstar.

GE Americom, formed from the original RCA satellite network called Satcom, is one of the biggest domsat operators and has seven satellites in service.

GTE owns and operates two separate satellite networks called Spacenet and GStar. GTE is a highly diversified company providing almost every option to its customers.

Hughes Communications Inc. (HCI) enjoys the status of being the world's largest commercial satellite network, following its acquisition of Western Union's satellite system called Westar, along with full marketing rights to the old Satellite Business Systems (SBS) network.

The HS 601 high power, wide bodied satellite was introduced five years ago to meet anticipated demand for a

Table 4.11 *American domestic satellites (domsats)*

Company	*Satellite fleet name*	*No of sats*	*Band*	*Comments on origin*
Hughes Communications	Galaxy	4	C band	
Inc. (HCI)	SBS	3	Ku band	
"	Weststar	1	C band	Acquired from Western Union
GTE	Spacenet/ASC		dual band	
GTE	G star	4	Ku band	
GE Americom	Satcom	4	C band	Formed when RCA bought by GE
GE Americom	Satcom/Alascom	3	Ku band	" "
AT&T	Telstar	3	C band	

Notes: 1 to the above total should be added a further three all Ku-band satellites for HCI, in orbit 1993

2 Alpha Lyracom is not included in the above table. Although a commercially-owned company in the USA it provides an international service in addition to US domestic traffic and is classified as an international carrier.

spacecraft with high power multiple payload, suitable for FSS and DBS (Table 4.12).

User services

Of the named private sector companies operating satellite networks, one of the more interesting is GTE Spacenet Corporation. It offers diversified satellite services to the business sector, entertainment, government, and the broadcasting industry. These specialist services are identified as:

1. Skystar Network Services. A two-way digital data communications service for organizations that rely upon centralized data processing.
2. Skystar Video Services. Includes broadcast business and television.
3. Skystar International. An all-digital private line satellite link.
4. News Express. A service to US television stations, allowing them to transmit live newsfeeds via satellite.
5. Systems design. GTE, a world leader in earth station design, engineers these for overseas customers. They have ranged in size from a 32 m C-band earth station down to the compact 1.8 m portable station.

Satellite Control monitors the performance of GTE's six satellites from its TTC stations located at Oxford, Connecticut, Woodbine, Maryland, San Ramon, California and Grand Junction, Colorado. This telemetry control is on a 24-hour day, year round, ensuring that the satellites do not deviate from their pre-arranged orbit. Overall control is exercised from GTE headquarters outside Washington DC, where the satellite control centre, network management centre and video control centre operations are housed.

PanAmSat: the world's first privately owned satellite

The monopoly held by Intelsat since the 1960s in international space communications was smashed when a giant Ariane 4 rocket roared skywards from its launching pad in Kourou, French Guiana, carrying with it a satellite called PanAmSat 1, now officially known as PAS-1. This was in the spring of 1988, and it was the result of the efforts of one person, Rene Anselmo. It was the culmination of his applications going back

Table 4.12 *HS 601 satellites ordered up to May 1993*

Customer authority	*Quantity*	*In service date*
AUSSAT	2	1 in service
SES ASTRA	3	Astra 1C, sched May 1993
American Mobile satellite corp	2	1994
Mexico Solidaridad	2	late 1993 and 1994
Hughes Communications Inc (HCI)	2	late 1994
Panamsat	3	1994 and 1995
US Navy	10	First one 1993
HCI for Galaxy network	3	Oct 1992, and 1993
ARABSAT	2	1995
Indonesia Palapa network	2	1996
Total to date	31	

to 1984 to Comsat for permission to compete for satellite communications traffic on the lucrative North American, Central American and South American routes.

The satellite PAS-1 has carried traffic on all of its transponders more or less continuously since the launch, which confirms the original views of analysts that the satellite business was assured of a very healthy growth rate well into the next century.

With international satellite traffic business earning in excess of 500 million US dollars every year, and with a future growth estimated at between 10% and 15% per year, there is every prospect that the satellite communications world will see more privately owned satellites like PAS-1.

With the opening up of Eastern Europe satellite operators have found new lucrative business. PAS-1 signed traffic agreements with the CIS and a number of its former Eastern Block countries. PAS-1 is capable of uplinking from 2.3 m dishes in the USA, and downlinking to dishes as small as 1.8 m in Eastern Europe.

PAS-1 transmits four footprints; the CONUS beam covering the continental USA, the north beam covering Central America, the Latin beam covering South America, and the European beam covering almost the whole of Europe. The maximum EIRPs for these four beams are as follows:

CONUS beam	45.5 dBW
Latin north beam	40.0 dBW
Latin south beam	38.5 dBW
European beam	47.5 dBW

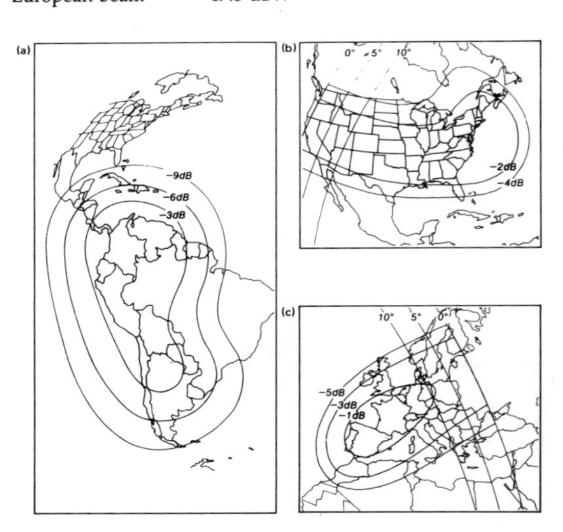

Figure 4.3 International coverage values (±1 dB). (a) Latin American beam has a 37 dBW beam centre. (b) US coverage provides a 45 dBW beam centre. (c) Pan American satellite covers Europe with a 47 dBW beam centre (Courtesy of Alpha Lyracom)

PAS-1 is a truly flexible satellite providing cross-strapping between Ku- to C-band, and C- to Ku-band. C-band transponders give out 8.5 watts of TWTA power and its Ku-band transponders give out 16 watts TWTA power. PAS-1 orbits at 45°W, and gateways to PAS-1 are provided through Alpha Lyracom's own earth terminal in South Florida. Alpha Lyracom has already signed contracts for three follow-up satellites, see Table 4.13.

These follow-on satellites are much more powerful, having a satellite launch weight of 6400 lbs against PAS-1's 2960 lbs, with a corresponding increase in output power from the transponders, 16 Ku-band giving 63 watts each, and 16 C band transponders with 30 watts each; transponder redundancy is provided at the ratio 10:8.

Table 4.13 *Alpha Lyracom's new spacecraft programme*

Satellite	*Ocean region*	*Orbital slot*	*Spacecraft type*	*Delivery*
PAS-2	Pacific	192/194 West	HS 601	Mar 1994
PAS-3	Atlantic	43 West	HS 601	Aug 1994
PAS-4	Indian	72/68 East	HS 601	Jan 1995

Newly formed national satellite systems

The Philippines is a nation of peoples fragmented by the geographical nature of the country. Although the island of Luzon contains the bulk of the 60 million Philippinos, there are citizens living on the 7000 islands and islets that make up the Philippines. The limitations of line-of-sight (LOS) transmission associated with UHF and VHF terrestrial broadcasting have meant that millions of Philippinos have never seen a television picture.

The Philippines has geographical characteristics much like those of Japan; hills, mountains, valleys and ravines are a natural obstacle to the reception of television signals broadcast by LOS means. To effectively cover the whole of the Philippines with VHF/UHF-TV would be an impossibility due to the several hundred thousand transponders and television stations that would be required, and even then this would not guarantee a 100% service.

The obvious solution to the problem is satellite broadcasting using a number of distribution outlets, and this is exactly what the Philippines government decided to do in April 1991. As part of a government modernization programme, Peoples' Television-4 (PTV-4) decided to implement satellite broadcasting using air time on Palapa-II, the Indonesian state-owned satellite. PTV-4 plans to simulcast its prime-time programmes initially in nine provincial stations on the island of Luzon, the Visayas region in central Philippines, and down to Mindanao. There are nine existing relay stations in the cities of Baguio, Naga, Cebu, Dumaqute, Bacolod, Davao, Zamboanga, Dipolog and Pagadian.

The second phase of the plan calls for an additional eight provincial points to link with the satellite feed or simulcast broadcasting by the end of 1992. These additional provinces are Laoag, Vigan, Ilagan and Tuguegarai in Luzon Island, and Puerto Proncesa in the Visayas region. Mindanao will have additional stations in Cagayan de Oro, General Santos and Cotobato City.

Still in the planning stage is a third expansion phase which, when completed, will result in an 80% coverage of Philippine's residential areas. The fourth phase, when satellite television reaches out to 90% of all Philippinos in all outlying islands, is still a long way off, perhaps 1998.

Asiasat

Since 1989 each passing year has seen more and more countries joining the exclusive club of satellite operators. One of the newest members of this club is Asiasat.

Asiasat is the Asian Satellite Telecommunications Corporation, set up by China's State Commission for Technology and Industry. In April 1990 the satellite Asiasat was successfully launched from a Long March 3 rocket from China's launching site. China is no newcomer when it comes to rocket technology; 26 successful and consecutive lift-offs have given China a remarkable track record in space technology.

The Asiasat satellite was built by the Hughes Aircraft Corporation, arguably the best in the world at building satellites for communications and broadcasting. China intends to launch at least nine more satellites before 1995 but whether it will use foreign-manufactured satellites or its own satellites is a matter for conjecture. China is well advanced in developing its satellite Dongfanghong III (The East is Red), which is expected to be available before 1995.

Dongfanghong III will transmit six colour television channels simultaneously along with 15 000 telephone, telegraph and data channels. A global scanning resource satellite is also being built as a co-operative project with Brazil, and it is known that China is well advanced with its manned space programme.

Asiasat will serve 30 countries with a total population of 2.5 billion. Sixteen of its total of 24 transponders were sold within four days after launch. Since then Asiasat has reported that its satellite business is taking off well. A contract was placed in October 1992 for a second satellite, Asiasat II.

Italsat

Another new member of the satellite club is Italy. In January 1990 the European Space Agency launched Italy's first satellite, called Italsat 1. This was launched from the same ELV that launched Eutelsat II F2, thus making a total satellite payload of 4263 kg. Both satellites were launched into a successful geostationary orbit. This development in rocket technology whereby more than one satellite can be carried by one rocket is an indication of the technological advances that are being made in space technology.

Hispasat

Spain is the most recent country in Europe to have the distinction of owning its own satellite system. This happened in December 1992 when satellite Hispasat 1A was placed into orbit from an Ariane 4 rocket. Some say this was Spain's way of celebrating the 500th anniversary of Columbus's discovery of the new world. Owning a satellite system has become the way for a nation to increase its prestige. It may be some time before Spain's satellite generates enough revenue to cover its initial cost but few experts doubt that eventually Hispasat 1A will be highly successful, for one thing it may cream off some of the satellite traffic to the Hispanic countries of Central and South America.

Hispasat 1A, technical data:

Manufacturer: Matra-Marconi
Transponders: Cover X-band, C-band, Ku-band
Services: FSS and DBS. The DBS service will provide TV coverage over mainland Spain and as far as the Canaries with a footprint from the 100 watt transponders of 56 dBW. Hispasat 1A will also provide an East–West link to America with a 36 mHz transponder.

Origins of direct satellite broadcasting for television, WARC 77

So far, most of the satellite communication systems described have been for the primary purpose of sending satellite signals from one fixed point on the earth to another. Services of this type are known as fixed satellite services (FSS). The earth stations used to receive these signals are relatively large, and professionally designed and engineered. Because of the dish size, which can often be up to 8 m, the sensitivity of the system is high. There are several different types of earth station, including main earth stations, secondary earth stations, portable earth stations, and now the so-called fly-away lightweight systems which can be transported by air anywhere in the world. Even so, the generalization still applies that all of these receiving systems are superior to any domestic satellite receiver.

Because of this fact, it is possible to operate FSS services for the transmission of speech, data or television with low output power from the satellite transponder. Powers used in transponders for FSS services can be between 8 and 15 W. Television signals transmitted this way are usually picked up on a medium-sized dish and re-broadcast, or sent down a cable television system.

The idea of using satellites for transmitting television signals direct to viewers in their homes was conceived many years ago but the technology of the day did not make this possible. This method of broadcasting was called DBS (direct broadcasting by satellite). The problems centred around making satellite receivers for a mass market, at a reasonable price and with the required sensitivity, and the building of satellite transponders with much greater output power, so as to guarantee a good grade of service to a receiving system inferior in performance to professional earth stations as described earlier.

The major difficulty was associated with the size of dish. The bigger the dish, the greater the gain, but it was considered impractical for domestic viewers to have large dishes which might prove unacceptable to many local authorities. It was thought that a dish size of the order of 50 cm was a reasonable specification.

Eventually the World Administrative Radio Conference WARC-BS took place in 1977 to lay down an international framework of frequency allocation for DBS. The world for this purpose was divided into three regions, as shown in Table 4.14.

Table 4.14

Region	*Area of the world*	*Frequency bands*
1	Europe, Middle East, Africa, CIS, Mongolia	11.75–12.5 GHz for downlinking
2	North, Central and South America, Greenland	11.7–12.1 GHz for downlinking
3	India, Iran, SE Asia, Australasia, Japan, China, Pacific regions	11.7–12.2 GHz for downlinking

The WARC Committee also laid down the basic fundamental requirements for any DBS system. These were: the television service should be available for 24 hours a day, it should not require complicated receiver systems, the dish size should be small, thus facilitating ease of mounting, and the risk of interference from other satellites and terrestrial transmitters must be kept to a minimum. Finally, the transmissions by DBS direct to viewers' homes must guarantee a minimum grade of service.

The WARC-BS 1977 recommendations to implement satellite television on a DBS basis (Tables 4.15, 4.16) posed a challenge to the industry in that the proposed new Ku band, 11–14 GHz, was a previously unexplored portion of the frequency spectrum which would require new techniques in

Table 4.15 *The WARC-BS 1977*
(a) Lower half 11.7–12.1 GHz

Channel group	*Frequency (GHz)*	*Orbit position*	*Country*	*Polarization R-hand or L-hand*
1	11.72748	19°W	France	R
5	11.80420	37°W	San Marino	R
9	11.88092	5°E	Turkey	R
13	11.95764			
17	12.03436			
2	11.74666	13°W	Ireland	R
6	11.82338	19°W	Germany	L
10	11.90010			
14	11.97682			
18	12.05354			
3	11.76584	37°W	Liechtenstein	R
7	11.84256	31°W	Portugal	L
11	11.91928	19°W	Luxembourg	R
15	11.99600	5°E	Greece	R
19	12.07272			
4	11.78502	37°W	Andorra	L
8	11.86174	31°W	United Kingdom	R
12	11.93846	19°W	Austria	L
16	12.01518			
20	12.09190			

Further assignments, all on orbit position 5°E, are as follows:

2, 6, 10	Finland	L
14, 18	Norway	L
4, 8	Sweden	L
12, 16, 20	Denmark	L

(b) Upper half 12.1–12.5 GHz

Channel group	*Frequency (GHz)*	*Orbit position*	*Country*	*Polarization R-hand or L-hand*
21	12.11108	37°W	Monaco	R
25	12.18780	31°W	Iceland 1	L
29	12.26452	19°W	Belgium	R
33	12.34124	5°E	Cyprus	R
37	12.41796			
22	12.13026	19°W	Switzerland	L
26	12.20698			
30	12.28370			
34	12.36042			
38	12.43714			
23	12.14944	37°W	Vatican	R
27	12.22616	31°W	Spain	L
31	12.30288	19°W	Netherlands	R
35	12.37960	5° E	Iceland 2	R
39	12.45632			
24	12.16862	19°W	Italy	L
28	12.24534			
32	12.32206			
36	12.39878			
40	12.47550			

Further assignments, all on orbit position 5°E, are as follows:

34	Sweden	L
38	Norway	L
24, 36	Nordic 1	L
22, 26	Nordic 2	L
28, 32	Nordic 3	L
30, 40	Nordic 4	L

Table 4.16 *Allocations associated with orbit positions at −31°, −25° and −19°*

Channels	*Orbit positions* −31° *R*	−31° *L*	−25° *R*	−25° *L*	−19° *R*	−19° *L*
1, 5, 9, 13, 17				LBX	F	
2, 6, 10, 14, 18	IRL	GNP	ALH	TGO	ZAI 2	D
3, 7, 11, 15, 19	LBR	POR + AZR		LBZ	LUX	BEN
4, 8, 12, 16, 20	G	CPV	ALI		ZAI 1	AUT
21, 25, 29, 33, 37	HVO	ISL		MRC	BEL	NMB
22, 26, 30, 34, 38		CTI	TUN		NIG	SUI
23, 27, 31, 35, 39	SRL	E + CNR	GHA		HOL	GNE
24, 28, 32, 36, 40				NGR		I

ALH, ALI	Algeria beams	I	Italy
AUT	Austria	IRL	Ireland
AZR	Azores	ISL	Iceland
BEL	Belgium	LBR	Liberia
BEN	Benin	LBX, LBZ	Libya
CNR	Canary Islands	LUX	Luxembourg
CPV	Cap Verde	MRC	Morocco
CTI	Ivory Coast	NGR	Niger
D	Germany	NIG	Nigeria
E	Spain	NMB	Namibia
F	France	POR	Portugal
G	Great Britain	SRL	Sierra Leone
GHA	Ghana	SUI	Switzerland
GNE	Equatorial Guinea	TGO	Togo
GNP	Guinea-Bissau	TUN	Tunisia
HOL	The Netherlands	ZAI 1, ZAI 2	Zaire beams
HVO	Burkina Faso (formerly Upper Volta)	R	right-hand polarization
		L	left-hand polarization

satellite receiver design. The satellite itself would demand higher powered transponders, thus putting emphasis on the development of more efficient solar power supplies. Higher powered transponders would call for a new generation of TWTs able to develop the required amount of power.

However, the benefits that would accrue from the introduction of DBS services more than justified the financial investment that would be required. These advantages may be summarized as:

1. A much wider choice in television programmes to the viewer; one satellite with 16 transponders would give a viewer the choice of up to 16 programmes.
2. Better picture quality. The use of the Ku band would make it possible to allocate many more television channels, and with a much greater bandwidth than those available in the UHF bands; the increased bandwidth of 27 MHz would make it possible to transmit high-definition pictures.

Non geosynchronous satellite systems

Though much of this book is concerned with geosynchronous satellites there are other types which are either non-synchronous or near synchronous. These are classified below:

(a) *Polar orbit* Launched at 90° to the equator so as to orbit over the poles; such a satellite will see a different region of the earth with each orbit, due to the earth's rotation.

(b) *Near-Polar orbit* These are launched at an inclined polar orbit.

Inclined orbit An otherwise geosynchronous satellite whose station keeping is restricted to East–West but is allowed to drift in the North–South plane so as to save fuel and extend the lifespan of the spacecraft.

(c) *Highly elliptical orbit* A special kind of orbit devised by the Russians. The Russian satellite *Molniya* uses such an orbit of 400 × 40 000 km, orbital period 12 hours, with an inclination of 63/65°. This orbital inclination was chosen so that the satellite's apogee remains over the northern hemisphere and able to provide coverage to every part of the old USSR, including the very far north, well inside the arctic circle. Such coverage would not be possible from geosynchronous satellites. To understand better the reason for the HEO one needs to examine the globe. The CIS extends from the Baltic states in W. Europe to the most eastern tip of Asia. The landmass extends over 11 time zones. In degrees longitude the CIS extends 172, almost half way round the world.

(d) *Low Earth orbit* These are discussed in the book. They may play more important roles for the future in MSS systems.

Operating inclined orbit satellites

When GSO satellites are nearing the end of the planned lifespan and fuel reserves are low it makes economic sense to let the satellites drift into inclined orbit by not using the north–south thrust motors. The drift rate makes it necessary to track the satellite with a moving dish at the earth station. In return the satellite operators sell transponder time at very low rates compared to the newer GSO satellites.

At the present time Intelsat has no less than four of its satellites operating in inclined orbit. Twice in every 24 hours these satellites drift alternately from north to south across the equator by as much as 4° latitude (Table 4.17).

In addition to these, all the Russian satellites of the Stationar Gorizont class operate in inclined orbit.

Table 4.17

Satellite	*Orbital slot*	*Max drift*
Intelsat V F5	66°E	3°
Intelsat V F7	57°E	2.3°
Intelsat V F2	21.5°W	4°
Intelsat V F4	31° W	3.5°

5 Launching into space: a risk industry

The economics of satellites and their use for communications and television broadcasting are complex. A number of main elements are involved in the system. These are:

1. The rocket, usually termed an ELV (expendable launch vehicle), distinguishing the technology from the reusable vehicle.
2. The satellite control system. This consists of ground stations that monitor and control the satellite during two phases: the launch phase and, secondly, the operational phase where telemetry data are received and processed in order to track and control the satellite in a predestined orbit.
3. The earth station with its uplink and downlink communications.
4. Finally, the satellite itself, which, if the launch programme goes as planned, will orbit the earth at an altitude of 35 786 km, travelling in the same direction as the earth, and with a speed of 11 700 km/h.

The important thing to remember is that all these are developing technologies, and, as is the case with any new and immature technology, development progress is, in the initial stages, very rapid: as the technology advances, further technical gains become that much more costly to bring about.

Rocket technology is a much older technology. The first real achievement was the V2 rocket produced in Germany in 1944, though no one knew at the time the satellite age began when the first experimental rocket V2 was launched from Peenemunde, Germany in 1944 with a trajectory of 118 × 30 miles high and a speed of 3500 mph. Had the war not finished when it did we might have seen the Sanger project come to fruition; this was a giant aircraft with a top speed of 11 000 mile/h intended for the purpose of bombing cities on the US mainland. The ideas of Eugen Sanger are still alive in the form of the Sanger II space aircraft. Sanger II is a proposal for a two-stage, winged space transportation weighing 400 tonnes. Unlike its ancestor this rocket is aimed at the conquest of space, and has a variety of uses, one of which will be as a launch vehicle for communication satellites. If Sanger II succeeds, its first flight will not take place until 2010. Sanger II is only one of the solutions to the ultimate plan which is to do away with the ELV altogether and replace it with a reusable spacecraft.

The expendable launch vehicle or rocket is an expensive item and it can only be used once. The Delta was one of the first launch vehicles for satellites. It had a length of 115 feet, weighed 116 tonnes and possessed a lift-off thrust of 305 000 pounds. Yet for all this power Delta's payload was 400 pounds for a geosynchronous orbit, or 5500 pounds for a near-earth

Table 5.1

Stage	*Burn time*	*Thrust (lb)*	*Fuel*
1	3.5 s	200 000	kerosene/liquid nitrogen
2	5.0 min	9 400	nitrogen tetroxide
3	45 s	9 500	solid type

orbit. Delta was a three-stage rocket with the characteristics shown in Table 5.1.

A more recent example of rocket technology and an indication of the strides made in this science is the Titan IV. This is a four stage rocket with the capability of placing a payload (the satellite) of more than 2500 kg in a geostationary orbit. Its characteristics for comparison with Delta are shown in Table 5.2. The Ariane IV, manufactured by Arianespace, is comparable with the Titan rocket although its payload may be greater. The top version of Ariane IV can lift a payload of over 4000 kg into geostationary orbit. However, in the future, all existing launch vehicles will be out-performed by Ariane V. This launch vehicle may not fly until the mid to late 1990s but when it does, it is expected to lift a payload of over 21 tonnes into low earth orbit. Its payload is expected to be the HERMES, a smaller version of the US space shuttle.

Table 5.2

Stage	*Burn time*	*Thrust (lb)*	*Fuel*
1	3 min	1.5 million	Solid type
2	1 min	550 000	Liquid type
3	2 min	100 000	Liquid type
4	3 min	30 000	Cryogenic motor

Failure risks

The life of the ELV (the rocket) is no more than six minutes or so and this assumes that all goes well, but for its payload, the satellite, life should be a good deal longer. Even so, the life of any satellite cannot be guaranteed. Like any other piece of electronic hardware, no satellite lasts for ever; and in the case of the satellite the risk of failure is everpresent.

The calculated life of a satellite can vary from 7 to 15 years. This theoretical life is dependent upon a number of factors, i.e. the design life of the HPA (usually a travelling wave tube), the fuel reserves for the transponder, usually dry batteries for powering the transponder during periods of eclipse, and the fuel reserves for supplying power to the satellite's motor, necessary to keep the satellite in its pre-assigned geostationary orbit.

With any piece of hardware, whether mechanical or electrical, there are two types of failure; these are defined as catastrophic or gradual wear-out. However, the effects of these failures are not always as might be expected. Catastrophic failure might occur in the first few hours of operational life but then again it might not occur until several thousand hours of life have been clocked. Even then the catastrophic failure may only affect one particular function of the equipment.

The important factor which distinguishes the communications satellite from any other piece of hardware or machinery is that once the satellite has been launched there is no possibility of replacing any faulty item, though some attempts have been made to recover satellites from space.

Hazards in space

Hazards in space fall into two categories; the possibility of a collision in space and the possibility of a malfunction or even catastrophic failure. Thirty years ago the possibility of a collision in space did not exist. Even today the risk of this happening is low, but in 10 years time it might be a different story. NASA has catalogued some 15 000 objects in space, some no bigger than a golf ball. Three thousand of these objects were man-launched and these constitute the biggest danger. Eventually they will break up or fall out of orbit but the danger exists.

In January 1978, and again in July 1981, Soviet satellites were destroyed by space collisions. In July 1983, less than two years later, the US space shuttle Challenger was struck by an unknown object which caused severe damage to a window.

Another source of damage or malfunction is that due to excessive radiation, e.g. proton radiation on the solar cells, and the effect of ultraviolet radiation on the spacecraft's external material. Yet another cause of failure in the spacecraft may be over-rigorous testing before the launch. By the same token too little pre-launch testing may not reveal the substandard components that might have found their way into the electronics compartments.

Other failures may be the result of faulty telemetry control, running out of fuel, software problems at the satellite TT and C station, or the launch vehicle running amok.

Of the first six Intelsat satellites launched, only four made it to attain geostationary orbit. This represented a failure rate of almost 50%. Yet, if we go back to the earliest origins of space exploration the CIS (then USSR) made three launches in 1957; all were successful. Was it exact engineering on a scale beyond the reach of any other nation, or was it luck?

The 1970s brought their share of major disasters when some vehicles blew up on the launching pad. In the 1980s failure was still a newsworthy item.

Satellites in space have a finite lifespan of between seven and 14 years at which stage the satellite's propellent for the orbital thrusters may expire. The satellite may then drift in space, constituting a danger to other satellites. It has now become standard practice to save sufficient fuel to kick the satellite out of its orbital slot and make room for its replacement satellite.

Economics of space communications

Notwithstanding the complexity of the modern communications spacecraft, the most expensive item in the overall chain of events is still that of getting the spacecraft into correct orbit. Like the early ballistic missile from which the modern launch vehicle has evolved, the launch vehicle is still a firework with a lifespan of no more than several minutes.

This writer spent several years in the 1950s in ballistics research on weapons, so what is stated about ballistic missile behaviour is not just conjecture. Most ordnance experts of the early cold war era would have been delighted if the intercontinental ballistic missiles (ICBM) had a 30% chance of reaching their target. The present day commercial launch vehicle is a much more reliable firework than the ICBMs but the risk of failure is everpresent.

Whilst the economics of building a communications spacecraft can be assessed, the cost of the launch is a grey area. A rocket or expendable launch vehicle (ELV) demands a launch capability. Gantries, fuel stores, fuel manufacturing plant, all adding up to the infrastructure of a medium sized town. In the USA, the CIS, and no doubt China, all of these facilities owe their existence to the Cold War. Even the technology itself owes its existence to the Cold War.

To summarize; without the defence expenditure, mainly by the USA, from the 1950s it is doubtful whether space communication would exist in the highly developed form to which it has evolved today. The true cost of making a space launch is difficult to assess, and is the result of 40 years of taxpayers' money. The commercial market for present day space launching is a highly competitive one, agencies seek the cheapest ride into space. For this reason, any country which chooses to subsidize its space launching capability will attract launch business, always providing it can guarantee a good success rate.

The most favoured contender for the launch market, apart from Arianespace which has captured over 50% of the world market, must be China.

More so than any other science, space technology is vulnerable to sudden and catastrophic failure: there are no soft landings. It takes thousands of scientists and engineers to place a spacecraft in orbit, yet all it takes for a diasaster might be one error. Thus the risk factor is related to the magnitude of failure rather than incidence of failure.

Space launching capability update

Today there are at least five nations with established launching capabilities for space exploration and space communications. However because of the overlap of interests between this application and military roles we are likely to see more and more countries joining this elite club within the next two or three years. Another factor playing an important role is national pride. India and North Korea, for example, are developing their launch industry.

In terms of numbers of commercial companies in this field, the USA still leads the world, and the number of players is expected to increase due to the run down in defence expenditure in the USA. Companies will have no alternative but to switch to the commercial market which is expected to grow at a very healthy rate at least for another decade.

Companies, or government led consortiums, and space agencies are listed in Table 5.3, with the country and launch vehicle series or rocket designation.

Table 5.3

Company	*Country*	*Launch vehicle series*
Arianespace	Headed by France	Ariane I, II, III, IV, with V under development
China Space Agency	Peoples' Republic of China	Long March rocket
US commercials		
General Dynamics Space Systems Div.	USA	Atlas rockets I, II, IIA and IIAS
McDonnell Douglas	USA	Delta rockets 6920, 6925, and replacements 7920, 7925
Martin Marietta Systems	USA	Titan rocket
National Space Development Agency	Japan	H1 and H2 commercial rockets
Soviet Space Ministry	CIS (previously USSR)	Zonda and Proton rockets SL12 and SL13

Source: Via Satellite, Vol III, no. 3 (1992)

Payload data for launch vehicles

Ariane (see Table 5.4)
Ariane IV is the most recent and the most popular of the Ariane series. The rocket is a three stage liquid propellant vehicle with a variety of strap-on boosters to meet different payload requirements. There are six possible configurations in increasing order of payload capacity. These are:

Ariane 40 No boosters
Ariane 42P Two solid boosters
Ariane 42L Two liquid boosters

Table 5.4 *Payload capacities for various missions and rocket configurations*

Mission	*Ariane 40* (kg)	*Ariane 42L* (kg)	*Ariane 44L* (kg)
GTO 7 degrees 200 km × 35 975 km	1900	3200	4200
LEO 5.2 degrees height 200 km	4800	7300	9600
LEO 5.2 degrees height 1000 km	3000	5200	7000
Polar 90 degrees height 200 km	3800	5900	7600

Source: Via Satellite, Vol III, no. 3 (1992)

Ariane 44P	Four solid boosters
Ariane 42LP	Two liquid and two solid boosters
Ariane 44L	Four liquid boosters.

Atlas (see Table 5.5)
Unlike Ariane vehicles, the Atlas first began life in 1957 as one of the first intercontinental ballistic missiles intended for reaching the CIS (then USSR). The commercial series was initiated in 1987 and is now available in four variants in increasing order of payload. Atlas II has lengthened boosters, Atlas IIA is a further enhancement on Atlas II whilst Atlas IIAS will have four Thiokal Castor IVA solid rocket motors.

Table 5.5 *Atlas payload performance for rocket types and missions*

Mission	*Atlas I* (kg)	*Atlas II* (kg)	*Atlas IIA* (kg)	*Atlas IIAS* (kg)
GTO 28.5 degrees 185 km × 35 786 km	2245	2676	2812	3493
Direct ascent, one centaur burn				
200 km	5800	6680	7000	8600
400 km	5200	5900	6200	7800
800 km	3000	3700	3900	4900

Source: Via Satellite, Vol III, no. 3 (1992)

Atlas IIAS has yet to fly a mission: its first flight will be for the Intelsat VII programme. It is possible to improve the Atlas payload capacity by launching into a subsynchronous elliptical transfer orbit and using the spacecraft to provide the additional energy by means of a perigee velocity augmentation (PVA) manoeuvre.

Delta (see Table 5.6)
The Delta II is the latest rocket version developed by McDonnell Douglas, and like the Atlas rocket described above, Delta II is a direct descendent of another ICBM, the Thor missile, first launched in 1960.

Over the next 25 years the Thor evolved into the Delta 3920 with a payload assist module (PAM) capable of inserting a 1720 kg payload into GTO. In 1984 NASA announced its intention to phase out the Delta in favour of the space shuttle but after the ill-fated shuttle blew up, work was once more

Table 5.6 *Delta payload performance for rocket types and missions*

Three stage mission	*Delta 6925*	*Delta 7925*
GTO 28.7 degrees 185 km × 35 786	1447	1819
GPS launch 55 degrees 20 184 km	850	1134
Two stage mission	*Delta 6920*	*7920*
LEO 28.7 degrees 185 km	3983	5039

Source: Via Satellite, Vol III, no. 3 (1992)

resumed on expendable launch vehicles (ELVs). Since then, the Delta in its variants has flown many missions, one of which was 20 Global positioning satellites (GPS).

Delta rockets are launched from Cape Canaveral at a latitude of 28.5 degrees for GTO launches, though some launches have been made from the Vandenberg Air base in California. Missions are either two or three stages, with the third stage being a Thiokal solid fuel rocket motor. Delta models 7920 and 7925 are improved versions with bigger payload capacities. The Delta possesses the ability to launch more than one payload into space and may be suitable for launching multiple small payloads into LEO.

The Delta in all its variants has proved to be a successful launch vehicle.

Titan

The Titan commercial launch vehicle, sometimes called Titan IV, is a descendant of another ICBM, one of the US Air Force defence weapons of the period from 1956 onwards. Titan I was an ICBM, Titan II was used by NASA for ten Gemini missions in 1965 and 1966, whilst the Titan III deployed more than 200 spacecraft of the Voyager and Viking type. Titan IV is a powerful combination of a liquid fuelled stage I, augmented by two solid rocket motors, and a liquid fuelled stage II. It has the capability of launching up to 5000 kg payload into GTO, or 14 000 kg into LEO.

H1 and H2 launch vehicles (see Table 5.7)

These are the launch vehicles developed by the National Space Development Agency of Japan, and unlike any of the launch vehicles built in the USA do not have defence origin. So far only the H1 model has flown missions: it was used for launching Japan's DBS satellites into orbit. The H1 is a three stage rocket with two liquid stages and the third being a composite fuel propellant. The H2 will be an improved version, having a far greater payload capacity.

Table 5.7 *H1 and H2 payload performance for different missions*

Mission	*H1*	*H2*
GTO 28.5 degrees	1100 kg	4000 kg
LEO 30 degrees 300 km	2900 kg	9200 kg

Source: Via Satellite, Vol. III, no. 3 (1992)

China (see Table 5.8)

China entered the commercial launch market several years ago, and since then it has competed favourably with the rest of the world launching companies. Its prices for a launch, coupled with the record of reliability it has acquired, have attracted business that would normally have gone to America or the French company Arianespace. All of the China rockets

Table 5.8 *Payload performance of China's Long March rocket*

Mission	*Rocket type*	*Boosters*	*Payload cap.*
GTO	LM 2E	4 strap-on liquid type	9200 kg

are identified as Long March type, with suffixes 1, 2, 3, 2E, and now Long March 4. To date China has successfully launched more than 30 communications or DBS satellites in orbit. Moreover it has had no serious accidents in its programme, except for the Optus 2 failure, but the launch phase was successful.

CIS (see Table 5.10)
The CIS (formerly the USSR) is another leading contender for the launch market. Although there have been some drastic cut backs in space defence expenditure there is no sign of a cut back in its commercial launch business which it has been trying to develop. In terms of launch vehicles, it is no secret that the CIS leads the world. As the CIS never went in for miniaturization its rocket technology has always been on a much greater scale than the USA. The rockets which the CIS is trying to promote for commercial launching of satellites is the giant Proton series, the SL12 and the SL13. These can lift massive payloads that make any western rocket look quite small.

Even more powerful than the Proton series is the Energiya SL17, designed to lift space stations into outer space. The SL12 has a mass weight of 699 tons and payload capacity of 5.4 tons, whilst the SL13 has an increased payload of 19.8 tons for roughly the same mass weight. The SL17 carries a massive payload of 153 tons for a mass weight of 2654 tons.

Launches in 1992
1992 was a successful year for the space industry. Arianespace took the record for the greatest number of launches (Table 5.9).

Table 5.9 *Major launches during 1992*

Launch company	*Launch rocket*	*No. of launches*
Arianespace	Ariane 4	11 major, plus some minors
McDonnell-Douglas	Delta rocket	11 major launches, mainly comms satellites for military use
General Dynamics	Atlas rocket	6 major launches, mainly DBS
China Great Wall Co.	Long March 2E	2 launches for Australia

Launching sites

Thirty-five years ago there were just two launching sites, one at Kazakistan in the CIS (then the USSR), and the other at Cape Canaveral, Florida. Both of these were built for the purpose of developing the giant ICBMs. As the defence space race marched hand in hand with space exploration, so more launching sites were constructed in America and the CIS. Today the Cold War may be over but the space race continues, this time with the developing countries; India, China, North Korea have joined the space club.

China has already begun construction of a further four launching sites at Jiuquian, Zichang, Taiyan, and Haian. India

Table 5.10

	Proton SL-13	*Energiya SL-17 (Four strap-on boosters)*
FIRST STAGE DATA		
Engine	RD-253	RD-170
Propellants	Nitrogen tetroxide/ UDMH	LOX/Kerosene
Thrust (tonnes)	167	806
Burn time (sec)	130	156
Specific impulse (sec)	316	336
Length (metres)	20.2	40
Diameter (metres)	7.4	4
Dry mass (tonnes)	43.4	25
Propellant mass (tonnes)	412.2	375
SECOND STAGE DATA		
Engine	(Kosberg Bureau)	(Not known)
Propellants	Nitrogen tetroxide/ UDMH	LOX/LH
Thrust (tonnes)	240	200
Burn time (sec)	208	470
Specific impulse (sec)	333	470
Length (metres)	13.7	60
Diameter (metres)	4.15	8
Dry mass (tonnes)	13.2	50
Propellant mass (tonnes)	152.4	821
THIRD STAGE DATA		
Engine	(Kosberg Bureau)	(Not known)
Propellants	Nitrogen tetroxide/ UDMH	(Not known)
Thrust (tonnes)	64	(Not known)
Burn time (sec)	254	(Not known)
Specific impulse (sec)	344	350
Length (metres)	6.4	(Not known)
Diameter (metres)	4.15	4
Dry mass (tonnes)	5.6	5
Propellant mass (tonnes)	47.5	25
PAYLOAD MASS (tonnes)	19.8	153
SHROUD MASS (tonnes)	3.0	–
LAUNCH MASS (tonnes)	697.1	2654
TOTAL LENGTH (metres)	59.8	60

Source: P. Clarke, *The Soviet Manned Space Programme*, Salamander Books

has two sites on Sriharikota Island in South India. Japan is constructing a new launching site on Tanegashima Island, Southern Japan from which it intends to launch its new H2 rocket in 1994.

Ideally a launch site should be situated on an island or in a remote location, surrounded by an ocean to permit aborted launches to drop harmlessly into the water. If the main purpose of the site is to launch spacecraft into equatorial geostationary orbits then it is preferable to locate the launch site as close as possible to the equator. This makes the launch more efficient and saves on fuel costs, it also means the rocket can carry a greater payload because of the smaller distance to the transfer orbit.

In the past Cape Canaveral was the most efficient launching site as well as the one that was known throughout the world. Today Kourou, in French Guiana, is the world's pace setter in both technology and its launching record.

The Kourou launching site, literally carved out from the swamps of French Guiana possesses all the sought after features. Located only 5° off the equator, Kourou has advantages that Cape Canaveral does not have; the benefits of east- and west-facing coastlines, and several thousand miles of splash-down accommodation in the Pacific Ocean.

Kourou, the commercial launch centre of the European Space Agency, is largely the brain child of France. It is an example of how the French government, acting in close co-operation with its high technology industries, is able to lead the world in many technologies. In theory the European Space Agency is an international consortium; however France has a 58.8% holding, Germany has a 19.2% holding, whilst Great Britain's holding is a mere 3.2%.

The space centre is completely industrialized. The site is criss-crossed with railway tracks so that one rocket can be rolled out from its hangar whilst the previous launch platform is being prepared for the next launch project.

As one pair of communication satellites are being prepared, fuelled up and assembled into the launch vehicle, the pair of satellites needed for the next space launch are undergoing system testing in the technical centre. Elsewhere in the technical centre, preparations are well advanced for the next generation of ELVs. This is the Ariane V which is due to roll out for its first test mission in 1995. Ariane V will carry a payload more than 50% greater than that carried by the Ariane IV vehicle. Because Ariane V is much bigger it requires a new style of gantry and umbilical mast.

Practically everything in Kourou has to be imported, but eventually the fuel will be manufactured and quality-controlled tested on site. A worthwhile feature of Kourou launch centre is the 60 m deep flame bucket filled with water from natural seepage from the swamp. Because of its unique shape this flame bucket also serves as a safety net should any employee fall from the high launch tower.

The launch

Whatever else has been written about space communications, the most exciting event is ignition and lift-off. Like any other firing every launch is unique, an experiment in physics which can never be repeated because of the variables of temperature, humidity, and barometric pressure. Even the fuel mixture will be slightly different, regardless of stringent quality control measures, but not measurably different from that in another launch vehicle of the same series. When lift-off occurs, and the rocket soars up to the skies it is the culmination of many weeks of pre-flight testing by space scientists and engineers.

For telecommunications and DBS, the most popular orbit for the spacecraft is the geostationary type of orbit, described elsewhere in this book. This is where the spacecraft is placed in an equatorial slot at a height of 36 000 km above the equator, travelling from west to east at approximately 3 km/s so that it appears as a stationary object viewed from earth. The technology for accomplishing this feat has been honed and refined since the first geosynchronous satellite Syncom was launched from Cape Canaveral in the early 1960s. Even so, the process of separating the spacecraft from its launch vehicle, and steering it towards its pre-destined orbit is always an exciting event, especially for the controllers on the ground at the TT&C station.

A geostationary launch is a controlled sequence of events, which each have to be precisely timed. The first is the initial launch into a parking orbit, an elliptical orbit bypassing the low earth orbit. This allows a short time to check telemetry, fuel reserves, and other control systems before moving into the transfer orbit zone, which is accomplished by the firing of the apogee motor on the space craft. This orbit has an apogee of the same value as the geosynchronous orbit so that another rocket boost at the right time, place, and with the correct amount of thrust, will place the spacecraft into its geosynchronous orbit. After this procedure, which occurs about 14 minutes from the ground launch, a number of events are initiated, adjusting roll, yaw, and pitch. These correct the effects of various gravitational forces acting upon the spacecraft which would otherwise cause the spacecraft to deviate from track. The main forces which affect the spacecraft are the sun, earth, moon and additional gravity variations due to the oblateness of the earth which bulges at certain points on the equator.

Assuming all goes well and according to plan, the spacecraft will maintain its predestined orbit to an accuracy of ±0.1° during the whole of its life in orbit. Drifts of up to 0.5° would affect reception from fixed satellite dishes. Keeping a space craft of the communications satellite variety on its correct track uses fuel and when this fuel is exhausted it is the end of the life of the satellite.

To extend this lifespan some satellite operators permit larger deviations than the standard of 0.1° from track and this

Table 5.11 *A typical flight record of a satellite launch*

Event	*Time into flight=T* (minutes/seconds)	
All systems checked at T minus 1 second		
Liftoff after ignition	0.00	
Booster engine cutoff	2.01	
Booster engine jettison	2.04	Phase 1 completed
Nose shroud jettison	2.30	
Second stage jettison	5.30	Phase 2 completed
Spacecraft separation	5.45	Phase 3 completed
Start programmed pitch positioning	10.00	
Stop pitch positioning	11.40	
Apogee kick motor (AKM) ignition on	13.45	
AKM burn-out	14.20	Phase 4 completed
Adjust forward velocity of satellite	14.25	
Start yaw positioning	14.45	
Stop yaw positioning	15.30	
Start roll adjustment	15.35	
Stop roll adjustment	15.45	
Blowdown hydrazine	19.30	
Deploy solar array	19.45	
Deploy search and rescue antenna	26.15	
Commence deployment of antenna systems	28.00	
Check control systems	33.45	
Handover of satellite to operator	34.00	Mission complete

Figure 5.1 Galaxy III launch from Delta rocket, September 1984 (Courtesy of Hughes)

deviation can be compensated for at the satellite earth station's tracking system.

Table 5.11 is a typical flight record for the launching of a satellite into geostationary orbit.

From available statistics of the commercial market (excluding military), Table 5.12 shows how the market has grown over the past 7 years.

Table 5.12

Year	*No of launches*	*Year*	*No of launches*
1988	15	1992	14
1989	12	1993	19*
1990	16	1994	24*
1991	16	1995	23*

* Taking account of planned launchings for 1993, 1994 and 1995, statistics show an increase of 24% over the first 4 year period 1988–1991.

6 Elements of satellite communication systems

The concept of satellite broadcasting held by the ordinary citizen is a simple and uncomplicated vision of a satellite in the sky that beams down television programmes, a television transmitter station in the sky. Reality is something quite different. The only thing that satellite broadcasting has in common with terrestrial broadcasting is the production studio and the programme link from the studio to the transmitter site. The basic elements in a satellite communication system are:

(a) the satellite
(b) tracking, telemetry and control stations
(c) various types of earth stations
(d) TVRO stations
(e) terrestrial re-transmitting stations

The satellite (a) may be regarded as being three separate units: the fuel systems, the satellite and telemetry control, and the transponder, which receives the uplink signal from the earth station and re-transmits this back to earth where it is received by (c), (d) and (e).

The TTC station (b) is probably the most important item in the system. This station takes control of the satellite during the critical period of launch. Thereafter its function is to hold the satellite on its track, making adjustments as necessary from time to time to correct for the effect of external forces such as solar radiation and gravitational forces.

Earth stations (c) come in various sizes, intended to function in different roles. They comprise a main earth station, a

Figure 6.1 Five-metre steerable dish for a TVRO installation in the Gulf region (Courtesy of Cable Television Services Ltd)

sub-earth station, and usually a number of smaller earth stations which might be mobile or transportable.

Finally, there is the terrestrial re-transmitting station (e). Perhaps the best example of a fully integrated satellite broadcasting system is that given by the BS-2 satellite broadcasting system operated by the Japan Broadcasting Authority, NHK, in conjunction with the Telecommunications Satellite Corporation of Japan (TSCJ) and the National Space Development Agency (NASDA).

This system, shown in Figure 6.2 is the most sophisticated system in the world. There are reasons for this. Firstly, it was Japan that pioneered the development of HDTV, which because of its bandwidth requirements – in excess of 30 MHz – could only be met by using gigahertz frequencies as used in satellite communications. Secondly, Japan has a harsh terrain, characterized by mountains, rivers and ravines, with populated islands separated by distances from the mainland beyond the range of UHF transmitters. It was therefore logical that Japan would be in the vanguard of satellite broadcasting of television programmes. The combination of satellite DBS as the backbone of the system, bringing satellite programmes to the bulk of the population, supplemented by re-transmission at UHF frequencies, meant that for the first time Japan had a television broadcasting system that was efficient, capable of broadcasting HDTV, and more important, capable of overcoming the transmission barriers, i.e. geographical features which prevented UHF from reaching thousands of viewers. Before the advent of satellite broadcasting in Japan, NHK had to invest in over 7000 television stations and over 6000 UHF translator stations, a far greater number of terrestrial transmitters than any other country in the world of a comparable size.

Earth stations

The giant C-band, standard A, B and C earth stations that were traditionally the gateway for international satellite communications from the late 1960s to the early 1970s have given way in the 1980s to the newer earth terminals with antenna dish sizes that range from 1.5 to 18 m. The basic form of any earth station is the same regardless of the system in which it is used. The send path consists of the interface with the baseband input signals, baseband processing, up-converter, high-power amplifier (HPA), usually a high-power klystron, the antenna feed and the final radiating element, the parabolic antenna dish.

The receive leg may be considered to be a mirror image, from the parabolic antenna, to the feed, low-noise amplifier (LNA), down-converter, baseband processing, and then the interface with the baseband output. Basic system parameters have been laid down for the different varieties of earth stations. The fundamental parameter of any earth station is the figure of merit, *G/T*, which is the ratio of the system gain to the overall noise temperature, where the system gain is the

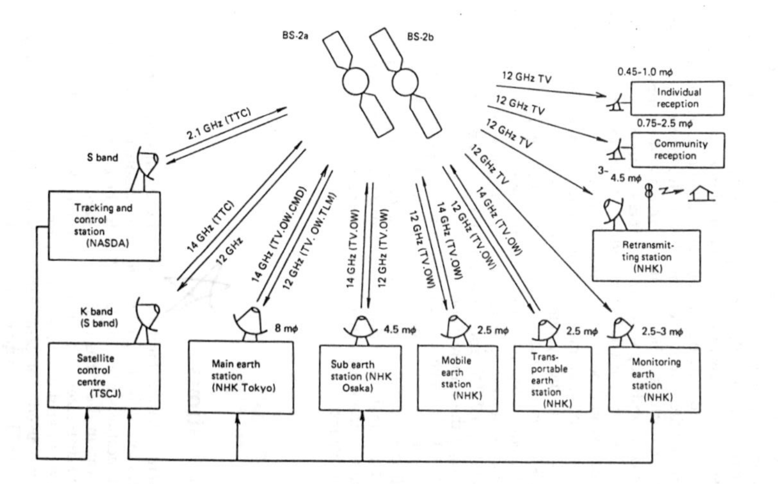

Figure 6.2 BS-2 satellite broadcasting system (Courtesy of IEE Publishing)

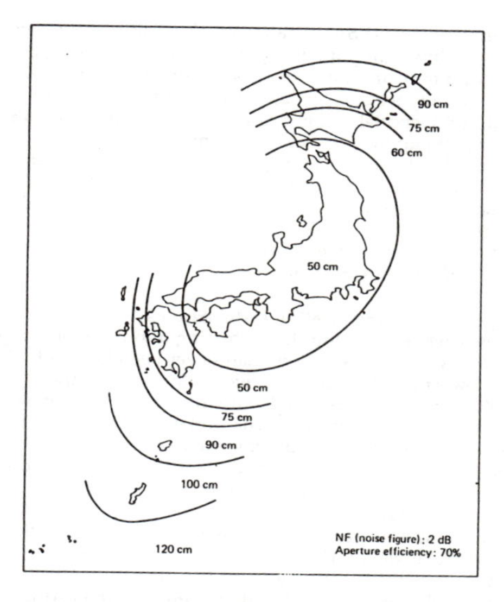

Figure 6.3 Japan: diameter of antenna required for DBS reception (Courtesy of IEE Publishing)

effective antenna gain, taking into account the actual antenna gain and the feed losses.

The front end of the earth station receiving system has to provide signal processing that will allow the recovered baseband signals to have a specific carrier to noise ratio *C/N* for the particular service the satellite system has to perform. This apart, the two most important features of any earth station may be considered to be the antenna and the output power from the HPA. The antenna must have a high gain in the forward direction with very low gain in other directions. Side lobe responses must be at the lowest possible figure so as to prevent unwanted radiation in the transmit mode, and to prevent unwanted pick-up of terrestrial-borne interference in the receive mode.

The antenna dish must maintain its specified performance figure under all environmental conditions and its pointing accuracy should be kept to specified limits, usually a fraction of a degree in elevation and azimuth.

The physical construction of dishes, whether used for transmission or receiving, is important. Poorly designed supports and front feeds can produce what is termed 'aperture blockage' as well as increasing the noise in the system. These unwanted effects can often be minimized by paying attention to the structure design, and by the use of microwave-absorbent material.

The design of earth stations has progressed much, particularly within the past decade. In addition to the fixed earth station fulfilling strategic roles, there are very small aperture terminals (VSAT), transportable earth stations, and fly-away lightweight portable uplift stations with dishes as small as 1.8 m. Currently there is a performance gap between transportable types and the latter fly-away types that can be packed in transportable flight cases but by the mid-1990s the performance gap may be expected to narrow.

Further developments are expected to occur in hybrid-type earth stations able to accommodate C- band, Ku-band and Ka-band frequencies. As more and more Ku- and Ka-band satellites are put into service, station owners will find it cost-effective to build one hybrid earth station to avoid the otherwise double investment costs.

Transportable uplink satellite earth stations

'Live by satellite' is a term that is coming into common usage, yet 25 years ago the expression had never been heard of. One of the newest technologies to arrive on the scene is the portable earth station for satellite uplinks. Today it is one of the fastest-growing industries. The portable earth station has many uses. These include sporting events such as racing, conferences, social gatherings, and official and state functions. The latest use is war reporting. Reporting live from a battlefield was added to the list during the Gulf War. At least a dozen agencies were involved in news reporting. Satellite Information Services (SIS) is only one of these agencies.

In August 1990 SIS sent a team of technicians with a portable earth station to Jordan. The team, and the earth station, travelled by a regular passenger flight to Amman, something that would have been quite impossible before the advent of the portable earth station. This earth station was transported in just five flight cases no bigger than many suitcases. Collectively these cases contained the subsystems. Three of the flight cases carried the antenna system and mountings. The first case contained an agile mount which allowed the dish to be manually steered to its correct azimuth bearing. The flexible waveguide was in the second case, whilst the third case contained the eight petal segments which bolted onto the rim of the drum. This system, when assembled, was an efficient 1.9 m parabolic dish antenna with a performance gain of 46.2 dBi in the Ku-band transmitting band 14.0–14.5 GHz, and 43.4 dBi in the receiving band 10.95–11.7 GHz.

The transmitting mode subsystem consisted of an up-converter, two 300 TWTs and a phase combiner. The RF signal was converted to Ku band, from 900–1800 MHz L band. It was then fed via a splitter to the dual 300 TWTs.

Each 300 W TWT produces an effective output of 23.7 dBW which can be configured by the phase and amplitude combiner to give 1 + standby, or as a phase combined pair whose resultant output of 600 W equals an effective output of 26.2 dBW. The output from the combiner is then fed to the antenna dish via a flexible waveguide. This transmitting subsystem is mounted on top of the antenna flight cases and acts as a counterbalance for the antenna subsystem when operational.

The ground communications subsystem, which comprises video and audio modulator and demodulator, provides the interface between the baseband video signal/audio signals, the

L-band signal applied to the amplifier and those returned from the low-noise converter (LNC). The modulator exciter accepts video/audio signals in any of the standard transmission formats PAL, SECAM, NTSC, and converts them into a modulated L-band signal suitable for input to the transmission subsystem.

The receiver subsystem of the transportable, fly-away earth station is housed inside the antenna mount. It consists of a 1 dB LNC and a filter network which produces an L-band signal for application to the demodulator.

The final part of the fly-away station is the audio/video monitoring and test subsystem. This produces video and audio monitoring, text signal generation, equalization, distribution, and output switching. Also in the flight case containing this test subsystem is a complete set of spares modules, including a spare video exciter and a spare satellite receiver.

The standard audio/video package deployed with the fly-away earth station is fitted for dual standards conversion PAL and NTSC. However, the satellite earth station can actually transmit in any of the standard transmission formats.

The complete earth station when fully assembled is powered up by 16 A from the standard electrical mains supply. This means that the earth station can operate from the usual supply found in most hotels throughout the world. This is a very important design feature, as it means that the installation could, if necessary, be used inside a hotel bedroom with the dish being mounted on the balcony or on the flat top roof.

Setting up a portable earth station is not a difficult task for a trained crew. Apart from the assembly, the essential items are a compass and an inclinometer. With these the satellite dish can be accurately positioned for azimuth and elevation.

Before operating a portable earth station in any country it is necessary to obtain regulatory approval from the appropriate telecommunications authority. The only exception to this international ruling is when the station is installed in an embassy building.

Initial on-air testing before transmission takes place usually consists of sending a test video signal in the prescribed format to the satellite network.

London based SISLink is an independent satellite uplink company formed in 1989 shortly after the UK Department of Trade and Industry granted its parent company Satellite Information Services a specialized satellite service operator's (SSO) licence. SISLink offers a range of satellite uplinking services and has particular expertise in the television broadcasting environment, regularly assembling and transmitting live television news (as in the Gulf War operations).

The SIS transportable earth station brief specification is as follows:

(a) System Advent Maris 1900
(b) 1.9 m antenna dish
(c) 2 × 300 W HPAs (phase combined)
(d) maximum EIRP 72.5 dBW
(e) test and monitoring
(f) own power source

Table 6.1 *Guide to SNG flyaway systems*

Company	*Product*	*Antenna diameter*	*Weight*	*Flight cases*	*Assembly time*	*Transmit EIRP*	*Receive G/T*	*Data capability*	*Options*	*Price*
Advent communications	Mantis 1900/1500 TV Flyaway	1.9 m 1.5 m	175–250 kg 158–233 kg	5 minimum 3 minimum	15 mins 15 mins	70–72 dBW 68.2–70.2 dBW	22 db/K 20 db/K	Yes Yes	Amplifiers up to 600 W, redundant amplifiers, redundant electronics packages, various encoding and transmission formats and data rates, remote control, audio, video, test and monitoring	£100,000-£200,000
	Mantis 1900/1500 Digital Radio Flyaway	1.9 m 1.5 m	175–250 kg 158–233 kg	4 minimum 3 minimum	15 mins 15 mins	57.1–72.5 dBW 55.3–70.2 dBW	22 dB/K 20 dB/K	Yes Yes	As above As above	£100,000-£200,000
	Mantis 1900/1500 Data Flyaway	1.9 m 1.5 m	175–250 kg 158–233 kg	4 minimum 3 minimum	15 mins 15 mins	57.1–72.5 dBW 55.2–70.2 dBW	22 dB/K 20 dB/K	Yes Yes	As above As above	£100,000-£200,000
Continental Microwave	SVX Ku-band Flyaway	1.5 m	76 kg	4	10 mins	72 dBW (500 W)	21.3dB/K	Yes	4 audios, TX frequency stability 2×10^{-8}, choice of RF spectrum filters to Eutelsat/Intelsat requirements	£100,000-£175,000

CPS Communications	Flyaway system	1.5 m	Typically 100 kg	500 W system 4 cases	15 mins	72 dBW (500 W)	50 dB/K	Optional	Motorised antenna system with remote control, test loop translator, redundant exciter system, power meter, frequency meter, audio/video terminal equipment	£123,000 £169,000
Dornier	SNG Fly Away Earth Station	1.5 m	236 kg	5/6	30 mins	69 dBW	20.5 dB/K	Yes	Ancillary communication Redundant/ combined HPAs	POA
Harris Allied	S1	1.8 m	488 kg	13	45 mins	73 dBW	23.9 dB/K	Yes	RF package	$210,000-$360,000
	S2	2.4 m	540 kg	15	60 mins	75.3 dBW	26.1 dB/K	Yes	RF package	$265,000-$375,000
Multipoint Communications	Ranger	1.5 m 2.4 m	225 kg	5	20 mins	69.5–78 dBW	23 dB/K 27 dB/K	Yes	1 × 300 W, 1 × 500 W, 2 × 300 W, 2 × 500 W, TV, Data, TV & Data, Ku or C-Band, tent	£100,000-£170,000

Source: International Broadcasting, EMAP Business Publishing

The manufacture of SNG transportable and fly-away systems is a healthy though competitive market. An SNG system usually consists of a packaged satellite system comprising the dish, exciters (usually there is a spare), communications receiver, high power amplifier (HPA), complete with its portable generator set.

SNG can either be of the portable type, termed fly-away, or packaged into a container mounted on a flat bed truck. Table 6.1 is a representative list of fly-away systems manufactured by six different companies in Europe and the United States.

Tracking of non-geosynchronous orbiting satellites

The ideal situation is one where once the satellite has been launched in a geosynchronous orbit, it should stay right there on track, maintaining its allocated position in orbit to a tolerance of 0.5°. This does not happen because external forces act upon the satellite to cause it to deviate.

The main forces on the satellite are gravitational due to the sun, earth and moon, with additional variable components due to the oblateness of the earth, which bulges at its equator at longitudes of 15°W and 165°E. Other forces include those due to the earth's magnetic field, solar pressure and micrometeorites. Once the satellite shifts out of orbit because of these forces, its orbit becomes inclined and the satellite is seen from earth as describing 'a figure of eight'. The total pull amounts to a shift of 1° per year, so not only does a satellite require an apogee motor to get back into orbit, but also station-keeping rudders and thruster motors.

Not all satellites operate in a geosynchronous-type orbit; some are launched in a polar orbit, or an inclined orbit. The CIS have a number of such satellites, two of which are Stationar 12 and Gorizont 3. These two operate nominally at 40°E in an inclined orbit. Gorizont 3 drifts at the rate of 1° every 38 minutes daily and appears as an elongated figure of eight moving above and below the polar arc.

Although most of the satellites for communications and broadcasting operate in a geostationary type of orbit in line with the earth's rotational axis, there are those which operate in an inclined orbit. There are also those which originally operated in a geostationary orbit but which have drifted out of orbit for various reasons. For satellites falling into these categories it is necessary for earth stations which require access to have steerable dishes in order to follow the drift of the satellite. Some of the satellites which require special tracking facilities include the Gorizont 3 satellites. These operate in an inclined orbit with a drift rate of 1° every 38 minutes daily and appear as an elongated figure-of-eight moving above and below the polar arc.

Tracking systems

Antenna tracking systems range from the simple single axis elevation tracking type to the sophisticated dual axis azimuth/declination type. There are essentially four basic types of tracking system which can be supplied and fitted to new antenna systems or in most cases can be retrofitted at a later date to fixed dishes:

(a) single axis elevation tracking
(b) single axis declination tracking
(c) dual axis azimuth and elevation tracking
(d) dual axis azimuth and declination tracking

Single axis elevation tracking This is achieved by the fitting of a linear actuator motor to the dish. Tracking in this mode is effective only in those cases where the longitudinal bearing is within 10° of that of the orbital location of the inclined orbit satellite; the elevation axis of the dish will then move perpendicularly to the polar arc, and hence tracking can be performed in the elevation mode only, provided that the beamwidth of the antenna is not restrictive. For example, tracking of the Gorizont 3 satellite at 40°E can be performed with a single axis elevation system.
Single axis declination tracking In this system the linear actuator motor moves the dish on its declination axis, i.e. in a diagonal direction. Tracking through the declination axis is effective at most receiving locations because the movement of the antenna will be perpendicular with respect to the polar arc and therefore the inclined orbital axis and the declination axis are identical.
Dual axis tracking systems These systems are preferable to single axis systems. They fall into two categories: azimuth and elevation, and azimuth and declination. The former is satisfactory where the location is less than 10° in longitude than the satellite's orbital position. Where the antenna location is greater than 10°, the azimuth/elevation system is less effective. This is because from such locations the inclined orbital track appears diagonal and tracking movements made in elevation and azimuth result in a step-type track, up and down and across.

The most efficient and universal in all roles is the dual track combination of elevation and declination. Such a system can track any inclined orbit satellite from any location irrespective of the longitude of the site. The linear actuator fitted to the declination adjustment point enables the dish to be moved in a perpendicular plane to the polar arc, thus tracing the line taken by the inclined orbit, and the addition of a polar mount actuator, or a horizon-to-horizon motor drive, enables the dish to be moved east or west to trace the apogee of the figure of eight.

This method of tracking is the most accurate since the track can copy and follow the inclined orbital track almost exactly, and can allow for any irregularities in the inclined orbit which can often occur.

Control systems for steerable dishes
Basically there are two methods: manual steering or automatic steering. The first is what it says; manual adjustment by an operator, the success achieved being dependent on the skills of the operator. The method has nothing to commend it, except for occasional use in fly-away-type dishes, and there is always the possibility of losing reception entirely if the manual movement is too great.

Automatic control is always preferable. In the simplest method the control signal is derived from the AGC signal level output from the satellite receiver; the controller, once programmed with the correct parameters and variables, will monitor the signal level and detect a reduction as the satellite moves away from its inclined orbit. This will trigger an optimization routine which will move the antenna in all directions very slightly until an increase in signal level registers. When the dish has finally homed into the satellite with maximum signal strength, the controlling system will shut down and wait for the next signal decrease to trigger another optimization routine.

There are some computer-assisted tracking systems that are so sophisticated that they can minimize actuator movements to 1/100 of a second. The AGC signal level detection can be set to register the slightest variation and trigger an optimization routine which uses as a reference the best signal level received, which has been recorded in the memory.

Some tracking systems employ computer terminals and VDU screens which provide visual data and maps to assist operator plotting of the tracking.

Wobbly satellites

This is the term given to satellites which deviate from their assigned track. Sometimes this deviation is erratic, and in other cases it is a gradual drift, but in either case it causes problems to reception. Sometimes the phenomenon lasts for a short time whilst satellite control seeks to regain operational control of the satellite, but in other cases it can be a permanent occurrence. Paradoxically, when this occurs it is a sign of a good satellite which has endured many years of reliable service but which eventually must come to an end when the fuel for the rocket motors is finally exhausted.

Keeping a satellite in its assigned geostationary orbit to an accuracy of 0.1° necessitates using fuel. However, the supply of fuel can be extended by permiting the satellite to drift out of this range. This practice is permissible within certain limits. Today, with many of the early satellites coming to the end of their life, more and more operators are seeking to get as much life out of their satellites as they possibly can.

Since Comsat was the first international satellite organization in the world, it has several of its early satellites still in operational service. The technique of keeping old satellites in orbit has come to be called the 'Comsat Maneuver', using attitude control to increase beyond the normal allowable 0.1° error. Earth stations follow the drift of the satellite, making occasional 'tweeking adjustments' to the earth station antenna.

With more and more satellites coming to the end of a long life, it is expected that the phenomenon may become more prevalent. However, we are not likely to see it occurring on DBS satellites for several more years.

Spacecraft and satellite manufacturers

The design and construction of spacecraft is a high technology, high risk industry where the rewards are not always commensurate with the risk. It is a technology that has evolved from the building of aircraft which explains the presence of major aircraft companies like Hughes and McDonnell-Douglas. As with any new technology which is burgeoning, other companies seek a share of the market with the inevitable result of overcapacity.

Up until a couple of years ago there were a dozen or more companies in the space industry. With a market not expanding as hoped companies merged. In the US the market leaders are Hughes Aircraft Company, GE Astro Space followed by Space Systems/Loral, formerly Ford Aerospace, whilst in Europe there has been a similar contraction. Alenia DASA, Aerospatiale and Alcatel have merged with Loral to make a super giant consortium. Matra merged with Marconi, and British Aerospace has offered its satellite business to both, but with no real positive result. In addition to these American and European consortiums and giant companies there are nascent satellite industries emerging in Asia. Satellites are becoming increasingly more costly, more sophisticated and more powerful, as typified by the HS 601 built by Hughes, and the GE Astro Series 700. GE Astro Space has a range of satellites that extend from Series 3000, the work horse of the first generation, through to Series 4000, second generation, Series 5000, and the latest, the Series 7000, the biggest and most powerful.

Hughes has a similar series. Its HS 376, the first of which was launched in 1983 became the most popular satellite ever built: 34 have been launched without a single hitch. The system parameters of this satellite are given in Table 6.2 and Table 6.6. The HS 601 is the biggest and the most powerful of the Hughes family of satellites. The best proof of its success rate is that more than two dozen have so far been built, with more orders going through to 1996. Evolution in satellite design and performance is illustrated in the comparison between the HS 376 and the HS 601 in Table 6.2.

The latest merger to take place in the satellite industry is GE Astro-Space and Martin Marietta Systems. On 1 April 1993 GE Astro-Space merged its business with Martin Marietta Systems (manufacturer of the TITAN rocket). The new company is called Martin Marietta Astro-Space.

Table 6.2 *Comparison between the HS 376 and the HS 601*

	HS 376	HS 601
Lift-off weight	2750 lbs	6400 lbs
Design life	10 years	15 years
Transponder power	50 watts	120 watts
No. of transponders	12 all Ku	16 all Ku

Note: Transponder numbers and power are typical, exact configurations are dependent on individual customer requirement.

Descriptions of satellites

The Olympus satellite

The Olympus satellite, when first envisaged, was the most advanced of its kind. It proposed the employment of Ka- and Ku-band transponders on a single platform. It was also designed to provide a variety of services, ranging from DBS broadcasts to business services and VSAT networks. A unique feature of this satellite is the high-power transponders. The satellite is also one of the largest in service. Its characteristics are shown in Table 6.3.

Table 6.3 *Olympus characteristics*

Satellite type	Three-axis stabilized
Satellite weight	2420 kg
Dimensions overall	21 m × 1.75 m × 5.3 m
Solar panel dimensions	25.7 m
Orbital position	19.0 W
Operational life	10 years
Payload	
Ka band	3 transponders
Ku band	2 DBS 27-MHz bandwidth, 62 dBW
TWTA power outputs	
DBS 340 W	
Ka 30 W	
Ku 30 W	

The Marcopolo satellite

In 1986 the Independent Broadcasting Authority of Great Britain granted a franchise to the consortium, British Satellite Broadcasting (BSB) to operate DBS services in Great Britain on three of the five channels allocated to Great Britain by WARC 77. BSB went into receivership and subsequently its operations were taken over by Sky TV. Nevertheless, BSB has a place in satellite history because of its commitment to DBS using a satellite specially designed for DBS broadcasting and subsequently abandoned. In 1992 Marcopolo 1 was sold to Norwegian PTT and renamed as Thor 1.

The satellite was a Hughes 376, one of the most reliable satellites in service. It is a spin-stabilized design. The despun section contains the transponders together with the deployable antenna whilst the spun section contains the power unit, attitude control, and propulsion and thermal control subsystems.

Communication subsystem The DBS communication subsystem consists of three single-frequency conversion 17/12-GHz transponders. Each channel is 27 MHz wide and is capable of accommodating C-MAC, D-MAC or analogue/digital FDMA transmissions. The transponders feature an ALC (automatic level control) circuit which can compensate for up to 15 dB uplift fades. A total of 12 Hughes-manufactured 55 W TWTs are parallel connected in pairs to provide six 110 W RF output amplifiers. Three of these can operate simultaneously at full power (see Figure 6.5).

The TWTs have an M-type dispenser cathode, a ground commandable filament voltage step feature and an anode

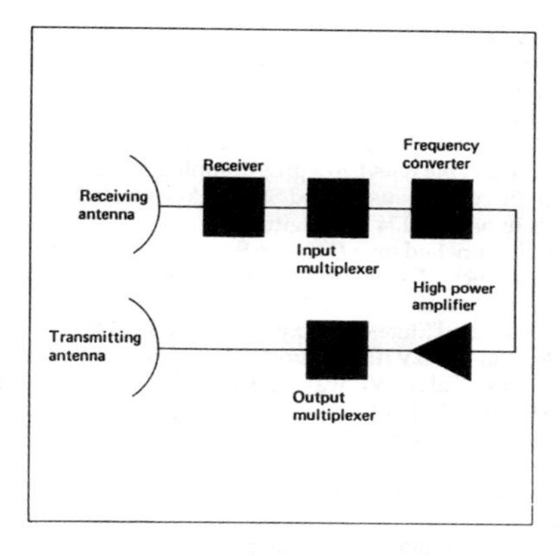

Figure 6.4 Simplified block diagram of a transponder (Courtesy of Thomson Tubes Electroniques)

voltage control loop whose purpose is to maintain a constant cathode current throughout the design life of the tube.

The six amplifiers are connected in a full ring redundancy arrangement, and provision is made to operate up to five channels at any time, provided that the total output does not exceed a total RF power of three channels operating at maximum power.

The communications antenna uses an elliptical reflector of 173 × 254 cm dimensions with a 14-element feed array (see Figure 6.6).

Service module The spinning Hughes 376 service module provides the correct thermal environment for the power, conditioning, and attitude/orbital control electronics. The solar cells are mounted on a deployable double cylinder and can generate almost 1000 W at end of life. Super nickel–cadmium batteries are used to provide sufficient power for normal service under eclipse conditions. A special design feature of these batteries is the ability to withstand high charge and discharge without electrode degradation. The system parameters are given in Table 6.4.

Table 6.4 *System parameters of the Hughes 376*

Lift-off weight	1250 kg
Spacecraft dry weight	490 kg
Diameter	216 cm
Height (on station)	752 cm
Service life	10 years
Array power	1024 W
Eclipse capability	100%
TWTA RF power	110 W (parallel operation)
Total number of TWTs	6 × 2
Operating channels and bandwidth	4, 8, 12 at 27 MHz wide
Minimum transmit EIRP	59 dBW
Minimum receive (*G/T*)	11 dB/K
Launch vehicle	Delta ELV (expendable launch vehicle)

The Galaxy satellite

The Galaxy satellites are powerful versions of the HS 376 communications satellite built by Hughes Space and Communications Group. The first Galaxy was launched on a Delta 3920 rocket on June 28, 1983. Dedicated to the distribution of cable television programming, it relays video signals in the contiguous United States, Alaska, and Hawaii. Galaxy I is in orbit at 134°W longitude.

Galaxy II, launched on a Delta on September 22, 1983, and Galaxy III, launched on a Delta on September 21, 1984, relay video, voice, data, and facsimile communications in the contiguous United States. Galaxy II is located at 74°W longitude, and Galaxy III is positioned at 93.5°W longitude. A fourth satellite, Galaxy VI, was launched from an Ariane 44L booster October 12, 1990. It is used for occasional video and newsgathering transmissions, and as an in-orbit backup for the five other C-band spacecraft in the Hughes Communications Galaxy/Westar fleet. It is positioned at 91°W longitude.

Each Galaxy satellite has 24 transponders. Instead of conventional leasing arrangements, Hughes Communications offered cable programmers the unique opportunity to purchase transponders on Galaxy I for the life of the satellite. The transponders on the Galaxy II and III satellites were offered for both sale and for lease.

Hughes Communications is one of 11 customers on five continents that have ordered the HS 376 spacecraft, the world's most widely purchased commercial communications satellite. Galaxy VI was the 32nd of this model to be launched. Based on technology pioneered by Hughes engineers, the HS 376 is spin-stabilized and has two telescoping cylindrical solar panels. A folding antenna also conserves valuable space in the launch vehicle.

Galaxy has a diameter of 7 feet, 1 inch (2.16 m) and is only 9 feet, 4 inches (2.84 m) high stowed in the launch vehicle. In orbit, the aft solar panel deploys, doubling the power output, and the antenna reflector erects for a combined height of 21 feet, 8 inches (6.6 m), or the equivalent of a two-storey building. The solar array of K7 and K4 3/4 cells, which generates 19.7 mW per square centimetre, produces 1089 watts of dc power at beginning of life; two nickel–cadmium batteries furnish power during solar eclipses. With its stationkeeping fuel, Galaxy weighs 1560 pounds (709 kg) at beginning of life.

The transmit and receive beams are created by a 6 foot (1.83 m) shared aperture grid antenna with two reflecting surfaces. One surface is sensitive to vertical polarization, the other to horizontal. Separate microwave feed networks are used for the different polarizations. Galaxy's average signal strength is 36 dBW, with each transponder using a multicollector travelling wave tube amplifier (TWTA) with a 10 watt output. There are six spare TWTAs. The satellites operate in the 6/4 GHz C band.

In the Delta launches, a McDonnell-Douglas payload assist module (PAM) was attached to the spacecraft to perform the conventional third-stage rocket function of insertion into elliptic transfer orbit. The apogee motor, which placed the satellite in near-synchronous orbit 22 300 miles above the equator, was a Thiokol Corporation Star 30 solid propellant rocket.

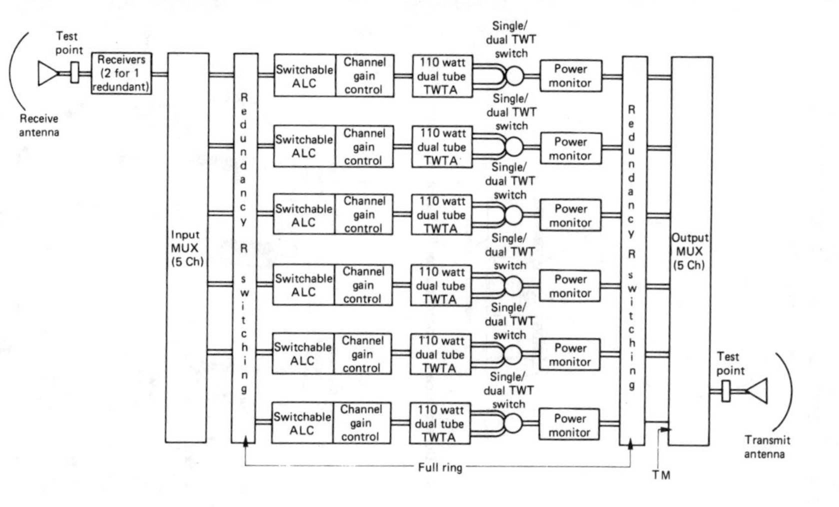

Figure 6.5 Functional block diagram of the communications subsystem (Courtesy of IEE Publishing)

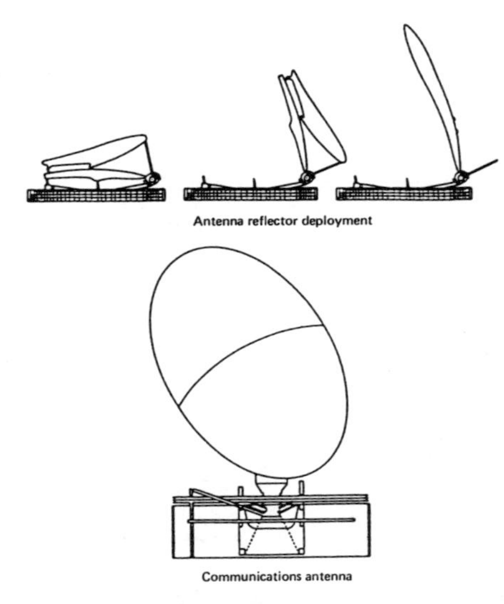

Figure 6.6 Communications antenna configuration (Courtesy of IEE Publishing)

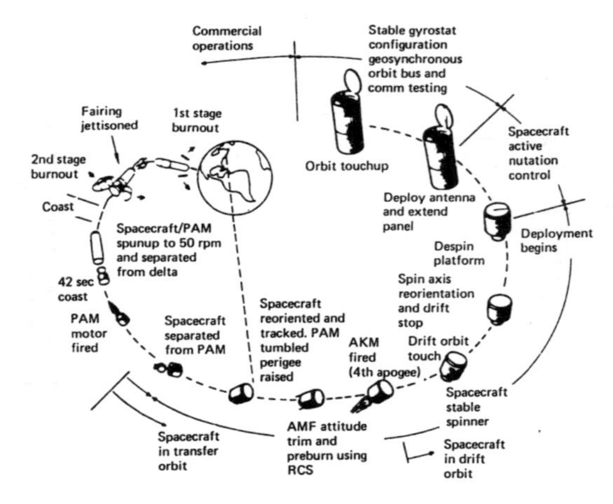

Figure 6.7 Launch mission sequence (Courtesy of Broadcast Engineering)

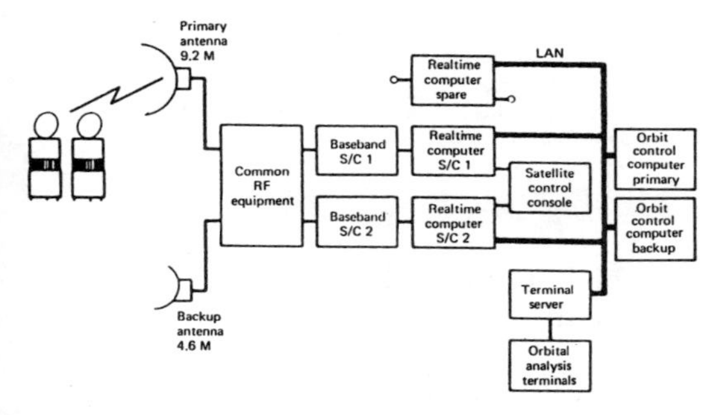

Figure 6.8 Satellite control facility system block diagram (Courtesy of IEE Publishing)

The Ariane booster injected the satellite directly into geosynchronous transfer orbit, so no PAM was required. The perigee was raised to geosynchronous altitude by a Star 30B solid propellant apogee motor.

The Transponder

Transponder is the term for the communication subsystem in the satellite. A satellite's communications payload can consist of many transponders which are multiplexed before being transmitted. A satellite's payload basically includes the receiving antenna which picks up the signals from the ground station, a broad-band receiver, an input multiplexer, a frequency converter for the receiving section in order to re-transmit the received signals on the downlink antenna

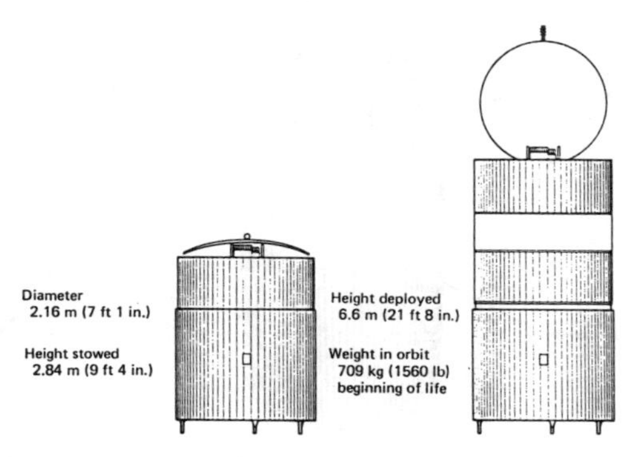

Figure 6.9 Galaxy Satellite HS 376

through a high power-amplifier, and a TWT with sufficient EIRP to be capable of being received on small receiving dishes in the case of DBS television programmes.

The C and Ku bands are used for telecommunications whilst the Ku band is used for DBS services.

How higher powered transponders have affected the bird

The communications capacity of satellites has increased with dramatic effect since the early 1960s. The capacity has increased two-fold, both in numbers of transponders and the power output per transponder. Transponders have increased from a small number on board to as many as 40, whilst output power has gone up from 8–9 watts on the first Intelsat spacecraft with 100 W now becoming commonplace. The constant upward trend in higher powered transponders is illustrated best in the Space Programmes Chart (Figure 6.11) provided by the French company Thomson Tubes Electroniques which supplies the TWTs for many of the space programmes. One of its latest orders is for a quantity of 34 TWTs type TH 3754, the latest generation of TWTs designed specifically for DBS.

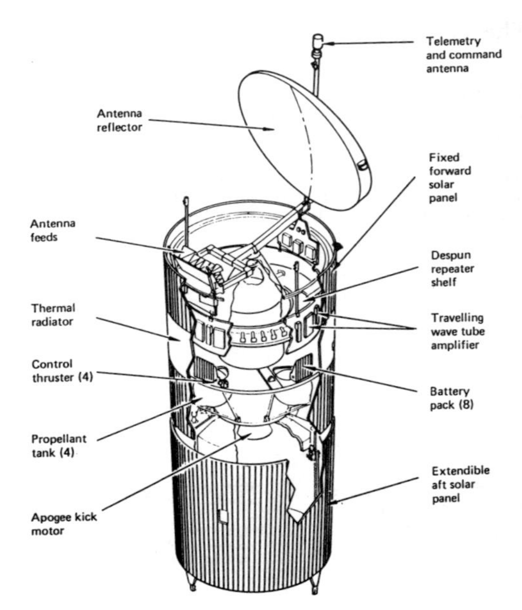

Figure 6.10 Galaxy Satellite HS 376

This upward trend in output power has placed heavy demands on satellite power supplies, and since the communications payload, i.e. the power amplifiers, represent about 75% of the total power consumed by the satellite, the rest being consumed by thrust motors etc., it follows that there is a need to maintain a high overall conversion efficiency of DC power consumed to RF power output. The weight of the payload mass is crucial, as any savings in satellite weight, both dry mass after launch and all-up weight before launch, make for a smaller launch vehicle and a cheaper launch.

Reducing the overall power consumption in the spacecraft resulting from more efficient power conversion means that corresponding savings can be made in batteries, heat ducting, cooling pipes, and the overall surface area of the solar wing surface. It has been calculated that a reduction in power consumption of only 10 W can bring about a reduction of the dry mass of 1 kg, and 2 kg of the launch mass.

The role of the electronic power converter in a satellite

The EPC (Figure 6.12) has the vital role in the bird of converting the spacecraft's low DC voltage bus rail into the very high voltages needed for the TWTs in the satellite's transponder. Reliability in these areas is crucial; if either the EPC or the TWT develops faults then the transponder can no longer function and the satellite becomes a useless object in space.

To operate a TWT requires high voltages at negative potential with respect to the satellite's body. Sometimes as many as six different EHT (extra high tension) rails are needed for the TWTs. The EPC is a DC to DC converter transforming the low-voltage bus rail to a well-stabilized EHT. EPCs are usually of the switched type where the input voltage is first chopped at the switching frequency (tens of kilohertz) in the power converter which is clock driven. The low AC voltage is then transformed up to high voltage, and rectified and filtered in the high-voltage section to supply the electrodes of the TWTs. The EPC also provides the necessary telemetry commands for the TWTs through which the correct operation of the transponder is checked.

Because the EPC is one of the heaviest components in a spacecraft and therefore has an important bearing on the total payload weight, it is important that it is no heavier than is necessary, whilst at the same time ensuring that its design and construction are such that it will withstand the *G* forces on initial launch. Since the EPC has to interface between the low-voltage rail and the high voltages for the TWT, whilst at the same time sending telemetry signals to the earth control station, the EPC may be considered to be a three-dimensional device with a transfer function of switching the TWTs from a non-operating mode to operating mode without generating unwanted transients.

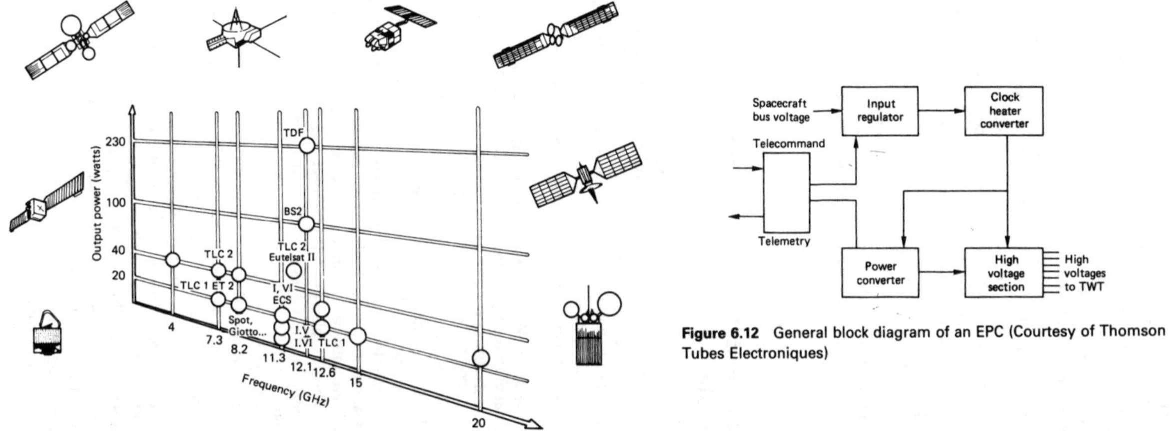

Figure 6.11 Space programmes: trends in output power (Courtesy of Thomson Tubes)

Figure 6.12 General block diagram of an EPC (Courtesy of Thomson Tubes Electroniques)

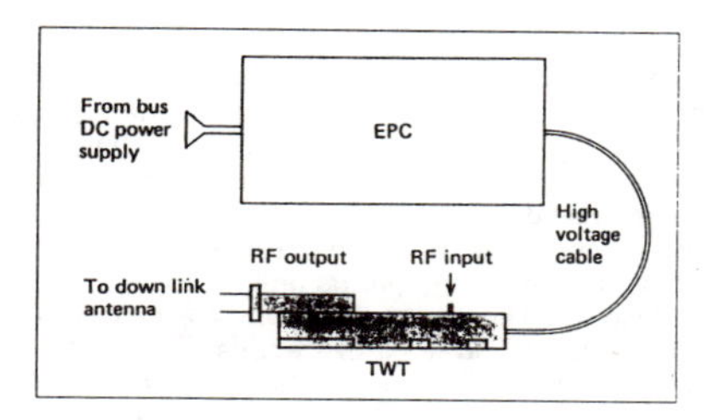

Figure 6.13 Schematic of a travelling wave tube amplifier (Courtesy of Thomson Tubes Electroniques)

Klystrons and TWTs

The power amplifiers used in the earth station and in the satellite itself are the heart of all satellite communication systems; they are the direct equivalent of the transmitters used in terrestrial broadcasting, with one main difference; the frequency bands for broadcasting.

At frequencies used for terrestrial broadcasting, they range from low frequency (LF) to very high frequency (VHF), whilst television broadcasting makes use of frequencies ranging from VHF up to UHF. At all these frequencies the conventional grid-modulated thermionic vacuum tube is quite capable of performing satisfactorily, though in general the power-handling capacity falls off with increasing frequency. The best example of this is the Thomson tube TH 539 which can generate 1.5 MW of carrier power in LF broadcasting, whilst at UHF frequencies the world's most powerful gridded tube generates 42 kW of peak sync power. But even this is a most remarkable achievement, having regard to the fundamental problems associated with the operation of grid-modulated tubes at UHF frequencies.

At these frequencies and beyond, two fundamental problems complicate the situation. These are:

(a) transit time
(b) capacitance effects

Both have a serious effect at UHF frequencies, and the magnitude of the problems increases with tube size. Transit time is where the flight time of electron flow from cathode to the control grid becomes a significant fraction of the wavelength of operation. At these low wavelengths the capacitance effects of the anode seriously retard the electrical efficiency. In the worst possible cases, both of these phenomena can actually destroy a tube, if the operating frequency is sufficiently high.

Fortunately, there is a solution to the problem and this is the velocity-modulated tube. With its operation based upon an entirely different principle, this type of tube excels at UHF and microwave frequencies. In this family of tubes it is the klystron that has achieved the greatest popularity. With an immunity to the problems mentioned earlier, the klystron has attained much importance in television broadcasting.

The klystron has both virtues and weaknesses, but most engineers are agreed that its virtues outweigh its weaknesses. Its greatest virtues are its reliability, strength and ruggedness, which endow it with a long life. Its only weakness is its low inherent efficiency. The klystron finds its greatest market in UHF-TV transmitters and in earth stations for satellite communications. In UHF-TV applications klystrons have been developed that can generate 70 kW of peak sync output power. In C-band and Ku-band applications the klystron can provide output powers of several kilowatts.

Although the klystron is a wideband device it has to have its cavities tuned to the frequency of operation, and thus the device has less attraction for applications in satellites. Klystrons therefore find their biggest application in earth stations where it is practical to retune the cavities of the klystron, should a frequency change be required. Almost all of the C-band earth stations use klystrons, usually with a duplicated arrangement.

In any high-power amplifier there are certain parameters which have to be designed to optimize performance. One of these is linearity. The klystron operates in a class A mode but when driven to saturation undesirable third-order harmonics can be produced. To prevent this from occurring it is usual for circuitry to be introduced to correct for the non-linear slope of the output curve.

Some examples of klystrons specifically designed for satellite earth stations are given in Table 6.5.

In 1990 Thomson Tubes Electroniques introduced its newest klystron intended for satellite uplinks. This is the TH 2456, a 14 GHz klystron with a minimum output power of 3 kW at saturation, and capable of being rapidly retuned to any of 6 or 12 preset channels depending on the tube model. It features simple permanent magnet confinement and uses forced-air cooling.

The TWT in the satellite performs exactly the same function as the klystron in the ground station, that is, it generates the RF power for the downlink, where the klystron in the earth station generates the RF power for the uplink. In terms of electrical efficiency there is virtually no difference. The main differences are in power handling and in bandwidth coverage. In general TWTs cannot generate anything like the power of a klystron but they have the advantage of greater bandwidth, a necessary feature on the spacecraft but dispensable at the earth station because the bandwidth can be changed.

As the TWT is one of the heaviest components in the spacecraft and because of the numbers used – up to 40 in some satellites – the development and research work has concentrated on reducing weight without any loss in performance. The TWT is required to provide the following service life:

(a) broadband operation
(b) high conversion efficiency (DC in to RF power out)
(c) low weight
(d) high linearity with a high gain

Table 6.5 *Examples of tubes used in earth stations and satellites*

Tube type	*Frequency band (GHz)*	*Output power*	*Overall efficiency (%)*	*Remarks and user*
Medium-power TWTs for satellites				
TH 3781	10.75–12.75	40–60 W	60	Telecoms 2
TH 3779	10.75–12.75	50 W	51	Eutelsat II
TH 3556	11.70–12.20	20 W	42	CTS/Hermes
TH 3626	12.50–12.75	30 W	40	Telecom 1
TH 3660	12.50–12.75	30 W	40	Olympus
High-power TWTs for satellites				
TH 3579	11.75–12.50	100–150 W	50	BS-2 Japan
TH 3619	11.75–12.50	200–230 W	50	TDF 1, TDF 2, L-Sat, Olympus
TH 3754	10.70–12.75	70–130 W	60	DBS applications generally
TH 3787	10.70–12.75	70–130 W	60	DBS second generation
TWT amplifiers for satellites				
TH 21754	10.70–12.75	70–130 W	51	Communications and DBS satellites
TH 21787	10.70–12.75	70–130 W	51	Communications and DBS satellites
TH 21619	11.70–12.50	200–230 W	42	Communications and DBS satellites
Klystrons for earth stations				
TH 2449	5.85–6.45	3 kW	38	
TH 2450	5.85–6.45	3.35 kW	38	
TH 2461	7.9–8.4	1.64 kW	38	
TH 2456	14.0–14.5	2.6–3 kW	25	
TH 2426	14.0–14.5	2 kW	25	
TH 2456	14.0–14.5	3 kW	25	

(e) ten or more years' life (the trend is upwards)
(f) high operating reliability
(g) rugged construction to withstand the *G* force of launch

Because the TWT is a thermionic tube with a hot cathode it has a wear-out mechanism and therefore the manufacturer tends to design for a longer life than the operational requirement, usually ten years. In a world where most of the functions of thermionic tubes have been replaced by solid state devices, it is worth noting that the TWT is the only device that can meet the requirements of the high-power satellite and it is precisely for this reason that so much research has been done on TWTs by Thomson.

The WARC recommendations for the introduction of DBS meant that the transmitted power had to be stepped up by a long way from the normal telecommunications satellites, which use powers of 9–20 W per transponder. In 1976 Thomson Tubes Electroniques initiated development of TWTs for DBS, aiming initially at the 100 W power level for use in the new Ku band, 11.7–12.5 GHz.

These new tubes marked a major milestone in space tube technology with the introduction of such features as pyrolitic graphite grid construction, pioneered originally by Thomson for shortwave broadcast tubes. This and other new features in tube construction enabled Thomson to develop the first 100 W tube, which went into service with the Japanese satellite BS-2B. Perhaps the most remarkable tribute that can be paid to the TWT is that after a satellite fell into the sea from a launch failure at a height of 9 000 m, when the TWT was eventually recovered and re-tested by Thomson engineers it met its original performance specification!

Major advances in tube development for space applications are as follows:

1969 Thomson commence development of 20-W 11-GHz TWT.
1976 World's first Ku-band TWT manufactured.
1977 Ku tubes in operational service.
1980 Start of the 100-W 11-GHz TWT development programme.
1986 100-W Ku-band tubes in service with Japan Broadcasting.
1989 A 58% efficiency is realized with Ku-band tubes.
1990 The magic 60% figure is realized.
1990 230-W tube in service.
1992 60% efficiency figure will be realized.

The latest generations of TWTs embody many new features which are essential for spaceflight applications in telecommunications and DBS applications, such as exceptionally high efficiency due to the use of multi-stage depressed collectors, long-life impregnated cathodes to ensure long tube life, and greatly improved RF performance characteristics to ensure mission performance in terms of a flat response versus frequency by as much as 15% of the centre frequency. This wideband response is necessary in DBS service because of the much wider bandwidth required to accommodate television signals.

Some of the current types of TWTs in service are shown in Table 6.5. The most powerful are the TH 3619 in operational service with the ESA Olympus, and satellites TDF 1 and TDF 2 owned and operated by France Telecom for the French state broadcasting TDF. Also impressive is the TH 3754 designed to operate in the 10.7–12.75 GHz band, and selected for use in DBS in the USA by the prime contractor Hughes Aircraft Company.

The modern TWTs not only possess an advanced performance which would have been impossible 20 years ago, but are also able to withstand the rigours of launch forces, and survive in space for as long as 17 years.

The basic elements comprising a TWT are the electron gun, the integral pole piece structure, and the final element, the collector envelope, which contains the multi-stage depressed collector (MSDC) which gives the TWT its high electrical conversion efficiency.

Much of the material in this chapter, on Travelling Wave Tubes, TWT, TWTAs and klystrons has been based on data from Thomson Tubes Electroniques. However this chapter would be incomplete without some reference to other world leaders in tube technology. In the field of space communications alone Siemens AG, Varian Associates and Hughes Electron Dynamics Division compete with Thomson.

In the sector of fully assembled solid state amplifiers for C- and Ku-band Lucas Aerospace Inc. is a major player, whilst in the business of building complete uplink terminals fitted with high power klystrons, Aydin Corp., Varian Associates and MCL Inc. tend to dominate the market.

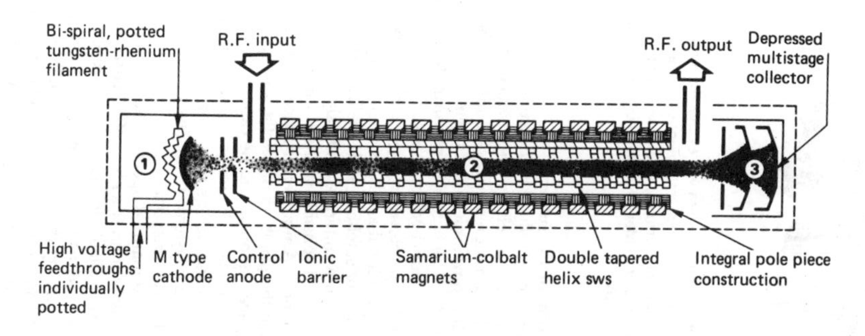

Figure 6.14 Space travelling wave tubes (Courtesy of Thomson Tubes)

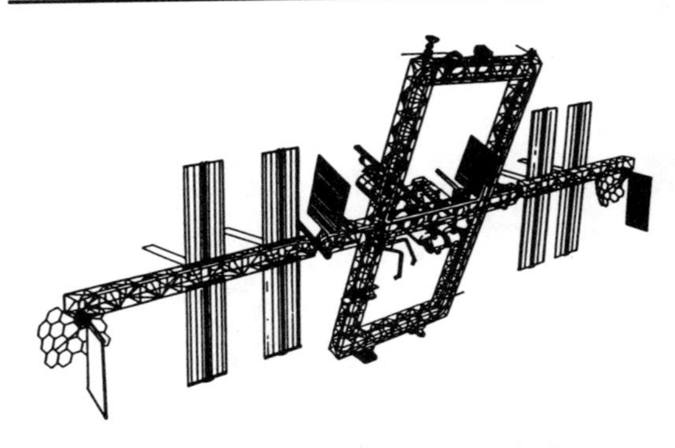

Figure 6.15 Space Station Freedom will evolve from a simple, unmanned platform into a permanently manned research facility, and then into this mammoth platform capable of servicing spacecraft on missions to the moon and to Mars. First-element launch date is set for 1995 (Courtesy of NASA)

The space platform of the future

Although we speak of satellite broadcasting, the satellite itself is merely the platform or vehicle upon which are mounted the components of the communication systems. As the demands for more powerful communication systems develop, so the satellite itself needs to be bigger in physical dimensions. As further satellites are launched into space, so we are hastening the day when collisions between these satellites become more commonplace. Operational satellites are not the only objects in space; there are faulty satellites and satellites whose lifespan has long expired still wandering aimlessly.

The only logical and sensible solution for the future is the manned space station. A number of space platforms would drastically reduce the numbers of satellites in space because each manned space station would be capable of housing as many as 100 transponders. A manned space station offers many other advantages such as refuelling in space, upgrading of the communication systems themselves, and repair and refurbishing of faulty transponders.

The concept of the manned space station is nothing new; what is different now from the ideas of 30 years ago is that the technology is now available to make the concept a reality. The major impetus for the manned space station came from military requirements by the USA and the CIS, but today the driving force is commercial as much as military. The GP or geostationary platform, as the project is now termed, is the subject of intensive studies by international space agencies such as NASA. Rockwell is the leader in this technology. The GP would be a massive modular structure, assembled in space. It would have crew quarters, workshops and transmitter equipment rooms, similar to those for terrestrial-based transmitters.

One alternative to the concept of the huge space station with overall dimensions of several hundred metres long is the satellite cluster idea. This would consist of a group of

6–12 GPs, spaced quite close together and linked together by optical inter-satellite communication links. Whatever solution is adopted in the future, the overriding problem is one of funding.

When the moon landings were curtailed nearly 20 years ago by the US administration, for a number of reasons, NASA wanted to embark on two projects: the space shuttle and the space station. However, the Nixon administration insisted that it choose one or the other. NASA opted for the space shuttle and shut down its work on space stations, whilst the Soviets intensified their study of earth station design, culminating in the Salyut space station. So, one way or another, when the concept of geostationary platforms for communication satellites becomes a reality it is almost certain to be based on Soviet technology.

7 Television transmission systems

Vision and perception

The eye is the visual communication system against which man-made visual communication systems are judged and evaluated. The eye sees a large amount of visual data simultaneously by virtue of the fact that it has several million communication channels acting in parallel at any instant in time. The electrical pulses that are generated by these millions of sensors in the eye are partly processed by the retina at the back of the eye and then further refined by the brain to provide the human experience of vision. The mass of detail forming this visual picture consists of two basic components: the detection of light (luminance) and the detection of colour (chrominance). There is also a third element called perspective, which gives the impression of depth to the picture.

No visual system made by man can match the performance of the eye, in speed or in quality of resolution. The process used in television cameras is electronic scanning. This is carried out in a manner similar to that in the eye, and the detail is translated into voltage variations that can be used to modulate the television transmitter. At the television receiver end, these signals are demodulated and used to vary the beam currents of the picture tube, the electronic beam of which is sweeping in synchronization with the transmitter scanning beam.

The fundamental weakness in the man-made viewing system compared with the human eye is the frame around the scene of vision. The human eye is under no such constraints. It can move in any direction, it can roam over a wide field of vision and it can track moving objects to a high standard of resolution. It can also re-scan an object.

Other qualities of the eye are paralleled in man-made systems. One such is 'persistence of vision'. When the eye registers a picture and when that picture is taken away the image is still retained by the eye for a short time afterwards, and the process of decay of the picture is an exponential decay. The cinema was the first industry to take advantage of this persistence of vision by presenting moving pictures as a series of individual frames per second. As the persistence of vision is something of the order of 1/10 of a second, it was found by the early cinema technicians that a projection rate of 12 frames per second would be sufficiently good to project a moving picture to the audience.

However, at this low picture projection rate, another factor comes into play, the awareness of the eye to interruptions in light. The eye has an acute awareness to light which increases with light intensity. Thus for projecting moving pictures with reasonable brightness it was found desirable to increase the picture rate. The eye has an awareness to a sudden interruption of light, and so it was found desirable to cut off the projection lamp whilst the picture was being moved on to the next frame.

The engineers who had the task of establishing standards for monochrome television standarized on the same picture aspect ratio as that for the cinema, i.e. 4:3. They also decided to use a picture rate in European television of 50 frames or pictures per second. In North America they chose a rate of 60 frames per second. The reason for these differences, which remain with us today, was simply the different standards in electricity supply systems, being 50 Hz for Europe and 60 Hz in North America and some other parts of the world, notably Japan.

Modern colour television is a development of monochrome television. Monochrome television required two pieces of information; the luminance component and the information for the two-dimensional picture. The addition of colour called for another component, chrominance. The ability of the human eye to perceive colour differences is less than its ability to perceive differences in brightness. In fact, there are wide differences between individual persons in the perception of different colours.

Television transmission systems. Formats PAL, SECAM, NTSC

Since the introduction of colour television broadcasting, first in the USA and then Europe, technical standards have been laid down. To a certain extent these standards were determined by the frequency of the electricity supply in the different continents. However, other differences emerged during the development of colour television broadcasting, with the unfortunate end result that there is no single international standard. This means that an NTSC television receiver produced for the USA market will not operate in Europe. Receivers designed on the PAL system in Great Britain will not work in France and so on. This chapter looks at these different standards, NTSC, PAL and SECAM, their individual and particular shortcomings. It also lists all the countries of the world with the particular television standard used. NTSC was introduced in the USA in 1963; PAL and SECAM standards followed in Europe in 1967.

International television standards

COLOUR STANDARDS

NTSC or NTSC 3.58 National Television Standards Committee
First used in 1954. R-Y and B-Y signals amplitude and phase modulate a suppressed carrier of frequency 3.58 MHz. Overall bandwidth of signals usually to 4.5 MHz maximum.
Disadvantages: phase errors cause hue, brightness and saturation errors.

Modified NTSC or NTSC 4.43
A term used by Sony for NTSC when replayed on European standard video cassette recorders. Although 525 lines 60 fields are replayed, colour subcarrier is shifted to 4.43 MHz. This enables the signal to be replayed on a PAL monitor which has loose-locked timebases and colour switch disabling.

PAL Phase Alternation Line
First used in 1967. R-Y and B-Y signals amplitude and phase modulate a suppressed carrier of frequency 4.43 MHz. Overall bandwidth of signals is 5.5 MHz in Great Britain, and 5 MHz in Europe. R-Y is reversed in phase every alternate line. A single delay line in the decoder enables the R-Y of two lines to be averaged, reducing phase errors.

PAL-M
This is a derivative of the normal PAL colour system with a shifted subcarrier of 3.58 MHz allowing it to be used on restricted vision bandwidth 525 line 60 field systems such as M, and 625 line 50 field system N.

SECAM Sequential Colour with Memory
First used in 1967. R-Y and B-Y are sent separately on alternate lines and used to phase modulate a non-suppressed subcarrier of 4.43 MHz. A single line delay enables the previous line's information to be decoded with the current line's. Tolerates large phase errors and is simple to record. Disadvantages: poor compatibility with monochrome receivers. Encoded SECAM cannot be vision mixed or faded using conventional techniques since full colour information is not available at a single moment in time. Either the studio must be designed as RGB with a single encoder after the mixer, or the encoded signal must be partially decoded into luminance and chrominance, mixed, and then re-encoded.

Vertical SECAM
The old system used by France, Russia, etc. Uses a truncated sawtooth added to the colour difference signals on lines 7 to 15 and 320 to 328 to indicate the actual sequence of the following chrominance information. Decoders for vertical are NOT compatible with those for horizontal SECAM.

Horizontal SECAM
The new system that omits the sawtooth and instead uses undeviated subcarrier on the back porch of each line to provide sequence information. Decoders are slightly more complicated but the horizontal system has the advantage of leaving field blanking clear for insertion test signals, Teletext, etc. Decoders for horizontal are compatible with vertical SECAM.

NOTE
The colour system shown in the following tables is that used for transmissions, but not necessarily origination. Many countries (such as Greece, Libya, Tunisia and Greek Cyprus) use PAL for studio origination (since SECAM is difficult to work with), but transcode to SECAM for transmission.

Standards Conversion and Transcoding

As television has originated in different countries, various technical standards have been developed, few of which are directly compatible. When each country originated its own programmes, this was not particularly important and those programmes that were exchanged were all film for which 35 mm and 16 mm are very simple standards that have not changed for over 50 years. While many programmes are still exchanged on film, live programmes can only be exchanged as video, and many programmes are now originated on video, rather than film. So it became necessary to develop equipment able to convert video pictures from one standard into another – the standards converter. Unfortunately, converting between different standards, particularly different field rates, is complicated and very expensive. Such standards converters cost around £100 000. These field store converters are required where 60 field NTSC pictures are converted into either 50 field PAL or SECAM, or vice versa. Converting between PAL and SECAM is rather easier and only requires a transcoder costing a few thousand pounds. On the other hand, if one simply wants to watch programmes from another standard, one can instead often use a multistandard video cassette recorder and monitor, although these do not allow copies of the programme to be made in the new standard. Below are listed the more common standards conversion problems, for the more important countries. This shows for each country the colour standard and television system. To determine the television standard, look up the system letter in the 'Television Systems' table (Table 7.1). Within a television system, there are the origination format (used with video) and the transmission format. Of importance to the origination format are the number of lines, fields, and colour subcarrier frequency, while the remainder are transmission characteristics which relate to transmitters, and television receivers or video cassette recorders with built-in tuners.

TYPICAL STANDARDS PROBLEMS

Britain (PAL System I)

Conversion: SECAM requires a colour transcoder (note if V or H SECAM is required); NTSC requires standards converter. Note that monochrome programmes are compatible in both PAL and SECAM areas.

Viewing: SECAM tapes may be viewed in monochrome on any PAL format video cassette recorder, and in colour on triple standard U-Matics when used with a multistandard monitor. NTSC will not replay on standard PAL video recorders, only multistandard types where the head drum speed can be increased to cope with the higher field rate, such as most U-Matics, except edit machines. When replayed, the NTSC becomes modified NTSC (or NTSC 4.43) with the subcarrier shifted from the normal 3.58 MHz to that used in European machines, 4.43 MHz. It may then be replayed on an NTSC 4.43 monitor (but not an ordinary NTSC set).

Transmission: System I (for UHF) is used by few other countries, although European BG is similar except for sound subcarrier (Great Britain 6 MHz, Europe 5.5 MHz).

North America, Japan (NTSC System M)
Conversion: PAL and SECAM require a standards converter for both monochrome and colour.
Viewing: a problem as there are as yet no video cassette recorders that replay true NTSC and PAL or SECAM. To replay either PAL or SECAM, a European video cassette recorder and multistandard monitor must be used. This may also include modified NTSC (NTSC 4.43).
Transmission: System M used for most NTSC transmissions, in USA influenced countries.

Western Europe (except France) (PAL System B and G)
Conversion: SECAM requires a colour transcoder: NTSC requires a standards converter. PAL B and G, and I are compatible on video tape, but not for transmission.
Viewing: as for Britain (PAL I).
Transmission: System B (for VHF) and G (for UHF) are widely used in Europe, Australia, Africa etc. Characteristics are similar to System I except for sound subcarrier which is lower at 5.5 MHz.

France (SECAM System L)
Conversion: PAL requires a colour transcoder; NTSC requires a standards converter. Note that because the phase relationship between luminance and chrominance is not as precise as PAL, transcoders provide one output that may be recorded but not mixed, and another for direct transmission, but which may not be recorded.
Viewing: PAL tapes will replay on SECAM video cassette recorders in monochrome, or in colour when used with a multistandard video cassette recorder and multistandard monitor. NTSC tape may also be replayed on a multistandard video cassette recorder, when used with a multistandard monitor, but will not replay in monochrome on a normal SECAM recorder.
Transmission: note that there are two different SECAM formats, Vertical and Horizontal (which are fully described under Colour Standards). Sets designed for Vertical (i.e. France) will not operate in the Middle East. SECAM is used on almost all different system standards, so compatibility between sets is difficult.

South America
This is a very difficult area, because Argentina, Paraguay, Uruguay and Brazil, all use standards totally different from the rest of the world. Argentina, Paraguay and Uruguay use PAL System N which is similar to normal PAL, but with the PAL subcarrier at 3.58 MHz so that narrower channels may be used. But since the line format is the same (625/50) as normal PAL, only a transcoder is required for conversion to or from PAL or SECAM. Conversion to or from NTSC requires a standards converter. Normal PAL standard video cassettes should however play on System N, in mono or colour. Brazil uses PAL System M which bears greater similarity to NTSC than PAL, with 525/60 format but a PAL subcarrier at 3.58 MHz. Conversion to or from NTSC requires a transcoder, while conversion to or from PAL or SECAM requires a standards converter. Only multistandard video cassette recorders will provide colour pictures from normal PAL,

SECAM or NTSC, although mono pictures should be obtained from NTSC tapes.

Table 7.1 *International colour television standards and transmission systems*

Country	*Standard*	*Transmission system*
Austria	PAL	B, G
Belgium	PAL	B, H
Canary Islands	PAL	B, G
Cyprus	PAL/SECAM	B, G
Denmark	PAL	B
Faroe Islands	PAL	B
Finland	PAL	B, G
France	SECAM	L
Germany (East)	SECAM	B, G
Germany (West)	PAL	B, G
Gibraltar	PAL	B
Greece	SECAM	B, G
Greenland	PAL	B
Iceland	PAL	B
Ireland	PAL	I
Italy	PAL	B, G
Luxembourg	PAL/SECAM	B, L, G
Madeira	PAL	B
Malta	PAL	B, H
Monaco	PAL/SECAM	E, L, G
Netherlands	PAL	B, G
Norway	PAL	B, G
Portugal	PAL	B, G
Spain	PAL	B, G
Sweden	PAL	B, G
Switzerland	PAL	B, G
United Kingdom	PAL	I
Eastern Bloc		
Albania	PAL	B, G
Bulgaria	SECAM	D, K
CIS	SECAM	D, K
Czechoslavakia	SECAM	D, K
Hungary	SECAM	D, K
Poland	SECAM	D, K
Romania	PAL	D, K
Yugoslavia	PAL	B, H
Africa		
Algeria	PAL	B
Azores	PAL	B
Benin	SECAM	K1
Burkina Faso	SECAM	K1
Congo	SECAM	D
Djibouti	SECAM	K1
Egypt	SECAM	B
Gabon	SECAM	K1
Ghana	PAL	B
Ivory Coast	SECAM	K1
Kenya	PAL	B
Liberia	PAL	B
Libya	SECAM	B
Madagascar	SECAM	K
Mauritius	SECAM	B
Morocco	SECAM	B
Mozambique	PAL	B
Niger	SECAM	K1
Nigeria	PAL	B, G
Reunion	SECAM	K1
Senegal	SECAM	K1
Seychelles	PAL	B
Sierra Leone	PAL	B
South Africa	PAL	I
Sudan	PAL	B

Table 7.1 *(continued)*

Country	*Standard*	*Transmission system*
Swaziland	PAL	B, G
Tanzania	PAL	B
Togo	SECAM	K
Tunisia	SECAM	B
Uganda	PAL	B
Zaire	SECAM	K1
Zambia	PAL	B
Zimbabwe	SECAM	B
Near and Middle East		
Afghanistan	PAL	B
Bahrain	PAL	B, G
India	PAL	B
Iran	SECAM	B
Iraq	SECAM	B
Israel	PAL	B, G
Jordan	PAL	B, G
Kuwait	PAL	B
Lebanon	SECAM	B, G
Oman	PAL	B, G
Pakistan	PAL	B
Qatar	PAL	B
Saudi Arabia	PAL/SECAM	B, G
Sri Lanka	PAL	B
Syria	SECAM	B
Turkey	PAL	B
UAE	PAL	B, G
Yemen (Arab Rep.)	PAL	B
Yemen (PDR)	NTSC	B
Far East		
Bangladesh	PAL	B
Brunei	PAL	B
Myanmar	NTSC	M
China (PR)	PAL	D
China (Rep.)	NTSC	M
Hong Kong	PAL	I
Indonesia	PAL	B
Japan	NTSC	M
Korea (Rep.)	NTSC	M
Laos	PAL	M
Malaysia	PAL	B
Philippines	NTSC	M
Sabah and Sarawak	PAL	B
Singapore	PAL	B
Thailand	PAL	B
Oceania		
Australia	PAL	B
Belau	NTSC	M
Guam	NTSC	M
Micronesia	NTSC	M
New Caledonia	SECAM	K1
New Zealand	PAL	B
Polynesia	SECAM	K1
Samoa	NTSC	M
Caribbean		
Antigua	NTSC	M
Antilles (Dutch)	NTSC	M
Bahamas	NTSC	M
Barbados	NTSC	M
Bermuda	NTSC	M
Cuba	NTSC	M
Dominican Republic	NTSC	M
Guadeloupe	SECAM	K1

Table 7.1 *(continued)*

Country	*Standard*	*Transmission system*
Haiti	NTSC	M
Jamaica	NTSC	M
Martinique	SECAM	K1
Montserrat	NTSC	M
St Kitts-Nevis	NTSC	M
Trinidad & Tobago	NTSC	M
North America		
Canada	NTSC	M
USA	NTSC	M
Central America		
Costa Rica	NTSC	M
El Salvador	NTSC	K
Guatemala	NTSC	M
Honduras	NTSC	M
Mexico	NTSC	M
Nicaragua	NTSC	M
Panama	NTSC	M
South America		
Argentina	PAL	N
Bolivia	NTSC	M, N
Brazil	PAL	M
Chile	NTSC	M
Colombia	NTSC	M
Ecuador	NTSC	M
French Guiana	SECAM	K1
Paraguay	PAL	N
Peru	NTSC	M
Surinam	NTSC	M
Uruguay	PAL	M
Venezuela	NTSC	N
		M

Note: All television systems listed employ an aspect ratio of the picture display (width/height) of 4/3, a scanning sequence from left to right and from top to bottom and an interlace ratio of 2/1, resulting in a picture (frame) frequency of half the field frequency. All systems are capable of operating independently of the power supply frequency.

System deficiencies in PAL, SECAM and NTSC

We all get accustomed to things, so in the long term we get more and more used to watching present-day television without being aware of system defects. The existing systems in use, NTSC in North America, and PAL and SECAM in Europe, all suffer from picture presentation defects to a lesser or greater extent. PAL and SECAM are generally accepted as giving better picture quality than NTSC but this is because these television standards came out after the NTSC standard. Some viewers become aware of these inherent defects more than other viewers but often put them down to poor reception.

In all three systems the colour information is sent with information on luminance (brightness), with the end result that we can get cross-luminance and cross-chrominance. When this occurs we get a situation where the black and white lines generate an irritating picture of dappled or zig-zag lines, and there is colour creepage from one part of the picture to another. Existing systems also suffer much from 'ringing', that is a sudden transition from black to white, causing an oscillatory effect.

The NTSC standard frequently results in colour swampage where the red colour predominates over other primary colours. Other defects common to all the systems are interline flicker, large-scale flicker and visibility of the line structure. Another common defect is the susceptibility to ghosting, where a second image is seen following on from the real image. Ghosting is caused by signals which are reflected from a large object such as a hotel or tower block.

One of the aims of HDTV is to eliminate all these defects, with the main objective to present a picture which is as near transparent as possible.

Picture quality: the quest for better resolution

The quest for communication has taken us from the crystal set age to the era of today. In television transmission we have progressed from the crude 25-line system of the 1920s to the colour presentation that is obtainable from the 26-inch television sets of today. Yet the pace of today's developments in this branch of technology is such that in another 40 years time a future generation of engineers might think of today's television receivers as being a crude attempt to produce television pictures.

A colour picture on a modern 10-inch television receiver looks pretty good, but see the same presentation on a 30-inch screen and viewing is less pleasurable, as deficiencies in picture quality become all too apparent. We see the horizontal lines and bright hues, patterning and dappling effects which are the unwanted effects of cross-luminance and cross-chrominance. This is an inherent defect in today's television receivers.

Apart from the defects described above, there are other deficiencies in today's television sets; in general, screens are too small, the aspect ratio is wrong, and the picture brightness on very large screens is usually insufficient.

The requirements for a high-definition system

There is no precise definition for high definition. The expression itself is relative. In the 1950s a 405-line monochrome television set was thought of as high definition, and this applied even more when the 625-line system was introduced in Europe.

Even when colour television arrived, screen sizes of average television sets were still relatively small and so the picture quality looked satisfactory. Yet even now, some twenty years later, television picture quality cannot be compared with cinema viewing as regards the technology. Better definition in television has been the goal of the broadcasting authorities around the world, but not until recent years has the technology become available, and the means to transmit HDTV to viewers.

Large screen presentation Broadcasting authorities are generally agreed that the television standards of today are not satisfactory for the future. HDTV implies a television screen presentation which is as near as is possible to the quality of the wide cinema screen. The cinema has made great strides in the development of wide screens and better picture quality, whilst television has stayed almost stationary. Much research has been done on the subject of HDTV, notably by NHK, the Broadcasting Authority of Japan.

One of the main subjects under study was screen size. The results of NHK's study was that optimum viewing is obtained when the viewing distance is $3H$ to $4H$ where H is the picture height. Expressed in practical terms, this means that a typical-sized Western European living room needs a television with a screen height of 30 inches (760 mm). Screen width is another important factor. Tests have shown that wide screen presentation corresponds better to what the human eye sees, and it can also contribute a feeling of a three-dimensional effect. Tests conducted by the BBC tend to confirm the results of NHK, but make the point that European living space is generally greater than that found in Japanese homes, and this factor itself reinforces the need for an early introduction of wide screen television viewing into Europe. There is now general agreement throughout Europe on what constitutes the ideal aspect ratio of a television screen; the figure is 16:9. For purposes of comparison, the existing television standard is 4:3. Where the pictures screened involve very considerable movement, viewing at close distances can cause eye fatigue. Other tests have shown that best viewing takes place at a distance of $3.3H$, and in the centre of the screen. As screens get bigger, so the human eye sees imperfections that are not visible on small screens. These manifest themselves as flicker, dots and lines.

Spatial–temporal frequency response of the eye For high brightness and contrast ratio, the human eye can resolve frequencies up to 60 Hz per degree spatially and 70 Hz temporally. At a viewing distance of $3H$ the television screen subtends an angle of 18.92° in the vertical plane. Thus by calculation if the television presentation is to be fully transparent it must faithfully reproduce spatial frequencies of up to 60×18.92, or 1135 cycles/picture height. Therefore, assuming that the camera and display act as ideal optical pre- and post-filters, then the television system must have at least 2270 active lines in order to satisfy the criteria.

The picture or field repetition rate is set by two factors. The first is the need to keep the level of flicker in the display at a low level, as the visibility of flicker increases with screen size, because of peripheral vision, and a higher field rate is therefore required. Experiments by broadcasting authorities indicate that field rates around 80 Hz are desirable. The second factor to take into account is the resolution obtainable from moving objects on the screen.

Under some circumstances the eye cannot perceive detail in moving objects because of the blurring caused by persistence. However, on occasions when the eye is following the motion of the object the image is effectively stationary on the retina, and hence the eye can resolve the detail to the same extent as when the object was not moving. Under these conditions, if the television camera is not itself following the moving object, blurring caused by the exposure time of each field can be perceived.

Experimental tests have shown that to reproduce moving objects in a fully transparent form a field rate of up to 1000 Hz may be required if motion blur is to be eliminated. These tests took account of the worst possible cases, which are not likely to occur all that often. Such a system might be described as attaining transparency. This is the term used to describe the almost perfect picture where it is impossible to distinguish between the picture before and after it has gone through the transmission system.

For near-perfect transparency a television system would, by calculation, require to have a standard of 2270 lines per second, and a field rate of not less than 80 Hz, if it was required to view at a distance of $3H$. Such a system would require a bandwidth of 350 MHz, and quite clearly this is an unrealistic target when it is remembered that the existing systems of PAL, SECAM and NTSC all use bandwidths in the VHF and UHF spectrum which are less than 6 MHz.

8 High-definition television systems (HDTV)

When 405-line transmissions were introduced in Europe, this was considered to be high definition, and so it was. This was primarily due to the fact that the screen sizes of most televisions were quite small, 8–12 inches across. As a result the system deficiencies were not so obvious; screen flicker and horizontal lines were barely perceptible. However, had it been possible then to build a television set with a 24-inch screen it would have been a different story. At that time in the early post-war years the television had not come to be perceived as a threat to the cinema, so there was no comparison between the two forms of entertainment.

When the first colour televisions came on the market after colour broadcasting commenced, the home television took on a new character; it now posed a threat to the cinema. The film industry, sensing this threat to its survival, started an intensive research and development programme to make cinema-going more attractive. It made some dramatic improvements in colour presentation, screen sizes and screen aspect ratios, finally culminating in the three-projector system fully synchronized to show an extra-wide screen. It also introduced some novel ideas for sound, so as to create a sense of realism for the viewer wherever he was seated. One of these was to reproduce sound effects on such a scale as to give the impression that the earthquake was actually taking place in the cinema.

All these improvements had the desired effect; viewers were quite happy to watch television for five days a week but when they wanted good entertainment on the grand scale there was nothing to touch the cinema. In the meantime, developments were taking place in television; screen sizes went up from 12 inch to 16 inch and then to 26 inch. Corrective circuits were devised to prevent picture flutter.

The new tubes were flatter faced and bigger, and used very high voltages which resulted in very bright picture quality with a wide range of contrast. All these improvements, added to the recently introduced 625-line system in the 1960s, gave television viewing a quality which it had never before had; it compared favourably to the cinema screen in viewing quality, albeit it was much smaller. Yet few viewers were conscious of the differences; they never realized or stopped to think that cinema viewing gave a far superior picture in terms of relative size. Seated in the centre of a cinema the viewer was seeing a picture size equal to a viewing distance of $3H$, where H was the height of the cinema screen.

To achieve the same kind of viewing experience on a television with a 9-inch-high screen, the viewer would have to be seated no more than three feet from the television. However, few viewers ever did this because the system deficiencies in picture resolution made viewing from such a short distance unpleasant. For television to reach the same degree of high quality, realism and range of colour contrast as the cinema, there needed to be some far-reaching improvements in both screen size and picture quality.

The scale of these improvements would have to be such that the existing systems of PAL, SECAM and NTSC, which had served the broadcast world so well for over 25 years, would no longer be acceptable.

High-definition television: by evolution or revolution

The different television transmission standards that were laid down in the 1960s were the first real attempt to bring about high-definition viewing. Competition is often the key to technical excellence. The USA introduced the NTSC standard, France developed the SECAM method of transmitting colour pictures, and Germany developed the PAL system. The NTSC system came first, with PAL (phase alternate line) and SECAM (sequential colour with memory) some several years later. The PAL and SECAM standards were an improvement over the NTSC method and became universally adopted throughout Europe and other parts of the world, with individual countries standardizing on one or the other. The result of this lack of acceptance of a common method of transmitting colour television had the unfortunate consequence that an NTSC television receiver could not receive transmissions from PAL or SECAM transmitters and vice versa.

It was this serious shortcoming that fuelled the need and subsequent development of television standards converters so as to enable 525/60-Hz field transmissions to be shown in Europe over the 625/50-Hz field for PAL and SECAM. There were other differences between the three systems; for example, PAL used FM for the sound channel, SECAM used AM, and NTSC used FM.

Now it looks as though history is about to repeat itself, with three HDTV standards to replace PAL, SECAM and NTSC. Japan has developed its MUSE standard, Europe has settled on the D-MAC family, whilst North America goes its own way with a standard of ADTV yet to be selected from a list of possible systems. However, with an eye to domination of the international market for HDTV, the Japan Broadcasting Authority NHK has developed the means whereby MUSE can be transmitted in the USA as a form of ADTV, whilst for Europe it has developed an HDTV to PAL and SECAM converter.

There is a growing feeling amongst broadcast engineers that a technological revolution in picture quality is long overdue; the 405-line system is now more than 50 years old and the 625-line system in Europe is now almost 20 years old. Developments in television cameras and in video recording technology have made giant strides during the same period when transmission technology has virtually remained static. Developments in very large scale integration (VLSI) technology have enabled equipment to be reduced in size by very large factors whilst at the same time improving performance specifications. The outstanding example of this is the VCR.

Thirty years ago video recorders were in the process of initial development but even when the first video recorders went into service in the 1960s they occupied several 19-inch racks. Today a VCR occupies a mere few panel inches of space.

Developments in the use of higher operating frequencies for television from band I to band III, then to UHF frequencies bands IV and V in the 1970s, and now the use of satellites, and satellite–cable delivery systems, reinforce the views expressed in the preceding section that the time is ripe for more fundamental improvements in television transmission systems which will bring the big cinema screen experience into the homes of the viewers.

Back in the early 1980s it was hoped that the world's broadcasting authorities would seek consensus on such a standard, but this did not happen. The stumbling block hinged on the question of whether progress should come by evolution or revolution.

Common to both approaches was the fundamental need for a greater bandwidth. The advent of satellites as a means of transmitting television pictures solved this problem. Because of the much higher frequencies used, in the C-band and the Ku-band, it now became possible to think about using higher bandwidth, as the Ku-band can easily support the use of a 40 MHz bandwidth.

The technology is here for HDTV but there are sceptics who believe that higher definition standards are not necessary. The argument rests on the claim that the public might like HDTV provided it is not going to involve capital outlay. The second point is that most of the viewers do not see any necessity for better definition because they see no obvious deficiencies with present systems. Proponents of HDTV argue that once HDTV sets are in the shops the product will sell.

What does seem an absolute certainty, in Europe at any rate, is that HDTV must come by evolution so that it does not make existing television receivers useless, and that the cost of any new HDTV receiver must not be significantly more than that of existing television receivers.

Implementation of HDTV in the USA

One of the first companies in the USA to begin research work into HDTV was the David Sarnoff Research Center. From this initial work the Center made the first over-the-air transmission of an HDTV signal on 20 April 1989 in New York. It was yet another milestone in the pursuit of technical excellence by the USA. This remarkable development in presenting HDTV took place when an HDTV signal was transmitted from television station WNBC at the World Trade Center in New York. The uniqueness of this demonstration lay in the fact that the transmission system, called advanced compatible television (ACTV), was fully compatible with the existing NTSC system. In other words, viewers in New York would have received the ACTV transmission on the ordinary NTSC television receivers, but without the wide screen presentation. This could only be received on the prototype ACTV receiver.

Since that first ever demonstration of an HDTV system in the USA, things have moved further ahead in the HDTV race. The fact that America had lagged behind Japan and Europe in any implementation of an HDTV system gave America a certain advantage in that, having not decided on a particular method of transmitting an HDTV picture, it could afford to consider all possible alternatives, including that of adopting an all-digital system in preference to the analogue systems proposed by Japan and Europe. Some experts are of the opinion that systems like the D-MAC systems adopted by European broadcasters will inevitably be replaced by a digital system within the not too distant future. Thus by adopting a digital system at this stage America might well overtake the rest of the world in HDTV technology.

In 1990 the Federal Communications Commission (FCC) invited proposals for a working system of HDTV with the additional requirement that the adopted system must be fully compatible with the existing 525/60 NTSC system. By the end of 1990 six American companies had announced their intentions of developing a digital HDTV system, thereby virtually assuring that whatever system the FCC finally adopted, it would almost certainly be all-digital.

At that time one of the foremost contenders was the Zenith Electronics Corporation, working in close co-operation with the research laboratories of the giant AT&T. A digital algorithm developed by Bell Laboratories would compress the HDTV signal, normally requiring a bandwidth of 27 MHz, down to a bandwidth of only 6 MHz without perceptible loss of resolution. The bandwidth of video signals can be compressed by digital techniques, something that would have been an impossibility less than ten years ago.

At one stage the FCC Advisory Board had before it more than 20 proposed HDTV systems; today there are only five. The list diminished as several proponents dropped out. Some merged their research work to form a consortium sponsoring one proposal. In the spring of 1991 there were six proposals put forward by six contenders. These companies were:

1. Zenith Electronics Corporation
2. General Instruments
3. Massachusetts Institute of Technology
4. Faroudjia Corporation
5. ATRC consortium
6. Japanese Broadcasting Corporation (NHK)

Since then the field has narrowed. Only one proposed system now has a sole proprietor, the NHK system proposed by NHK. Interestingly, this system, called narrow band MUSE, is more of an analogue system. A consortium composed of Zenith Electronics Corporation and AT&T is proposing a digital system called Digital Spectrum Compatible HDTV (DSC-HDTV).

The differences between the five proposals can best be explained by the numbers of scanning lines employed, the frame rate, and whether the system uses progressive scanning or uses interlacing. The present day NTSC system uses 525 lines per frame, and 59.94, commonly called 60, frames per second, and with a 2:1 interlace. In television standards language this is called a 525/60/2:1. By comparison the

Japanese MUSE system is a 1125/60/2:1, the ATVA interlaced system is a 1050/60/2:1, whilst the ATVA progressive runs at 787.5/60/1:1. DSC-HDTV operates at 787.5/60/1:1, ADTV uses 1050/60/2:1, and ACTV uses 525/60/1:1 (Table 8.1).

The choice of these numbers is a measure of the apparent picture quality of the system, but conversely the better the picture quality, the less likely is it that the system will blend in with the existing NTSC standard. The 1125/60 of narrow MUSE has compatibility with the 1125/60 production standard but the line and frame rates will require elaborate conversion equipment to convert to NTSC. On the other hand, whilst the proposed 1050 standards may lack the resolution of the 1125 system, being a direct multiple of 525 line it will be easier to convert.

An even more important factor in the introduction of HDTV to the USA is the fact that the FCC is insisting that any new HDTV system must be fully compatible with existing television receivers. It is almost certain that the solution will lie in simulcast. A simulcast system is one where the new HDTV service will be broadcast along with NTSC transmissions.

Table 8.1 *Competing systems for HDTV broadcasting in the USA*

Proposed system	*Lines/Hz*	*Scan*
ACTV: advanced compatible TV proposed by NBC/Philips/ Thomson and Sarnoff Research Center	525/59.94	1:1
Narrow MUSE proposed by Japan Broadcasting Authority NHK	1050/59.94	2:1
Digital spectrum compatible HDTV by Zenith Electronics and AT&T	787.5/59.94	1:1
Advanced digital ADTV proposed by NBC/Philips/Thomson/Sarnoff	1050/59.94	2:1
ATVA progressive system proposed by General Instruments and MIT	787.5/59.94	1:1

The FCC is insistent that any form of HDTV or ADTV adopted in the USA will be suitable for being broadcast over any terrestrial network using the 6 MHz channel bandwidth. It is also insistent that the same programme should be transmitted to viewers who have television receivers for NTSC format.

The Zenith proposal makes use of compression techniques which will compress the HDTV programmes in a 30 MHz bandwidth, down to a 6 MHz bandwidth suitable for transmission over the UHF network. At the receiver end the compressed bandwidth is expanded to a full 30 MHz and displayed on an HDTV receiver.

The FCC Testing Center will evaluate all five proposals in a test programme. Laboratory testing will take seven weeks, after which the systems deemed acceptable will be simulcast for a period of one to two years. After full evaluation by FCC experts the final decision will probably not be taken before 1994. All the five proposals will undergo identical evaluation tests for:

(a) *Interference performance* Vis-a-vis NTSC channels as well as channels carrying HDTV signal; co-channel, adjacent channel and taboo channel conditions
(b) *Susceptibility to impairments* In broadcast and cable transmission, impulse and random noise, multipath/microreflections, aircraft flutter, discrete frequencies representing other radio services, second and third order distortions, etc.
(c) *Image quality* Subjective assessments by typical viewers and detailed comments be expert viewers on static and dynamic resolution, impaired and unimpaired video quality, etc.
(d) *Audio performance* Objective measurements and subjective evaluations, both impaired and unimpaired.
(e) *Key operating information* Peak-to-average power, scene-cut and content transitions, service area fringe, performance, etc.

The odds are that the system finally selected by the FCC examining committee will be one of the four digital proposals, and that by 1995 American broadcasting will be leading the world with a high definition television system capable of being transmitted from satellite, terrestrial transmitters or cable.

Manufacturers of HDTV transmission equipment

When it comes to putting new technology into production the USA leads the rest of the world with its speed of response. Already the major manufacturers are gearing up. On 25 January 1993 Harris Corporation issued a press release to say it will 'develop and manufacture broadcast equipment for the Zenith/AT&T Digital HDTV System. Harris will support the rapid deployment of digital four-level vestigial Sideband modulation and transmitting equipment to broadcasters'.

Harris will licence the digital signal compression HDTV (DSC-HDTV) if the FCC adopts the Zenith/AT&T System. This technology would be used in exciters for Harris HDTV terrestrial TV transmitters. HDTV transmission processing equipment combines and formats digitally compressed video, compressed audio and ancillary data for error-correction, pre-coding and data equalization so as to remove transmission errors and ghosting. Rejection of other interference from any existing NTSC transmissions is also part of the transmission processing and modulation equipment.

Advanced definition television (ADTV) in the USA

HDTV broadcasting and DBS are synonymous in Europe, and also in Japan, but this is not the case in the USA and there are

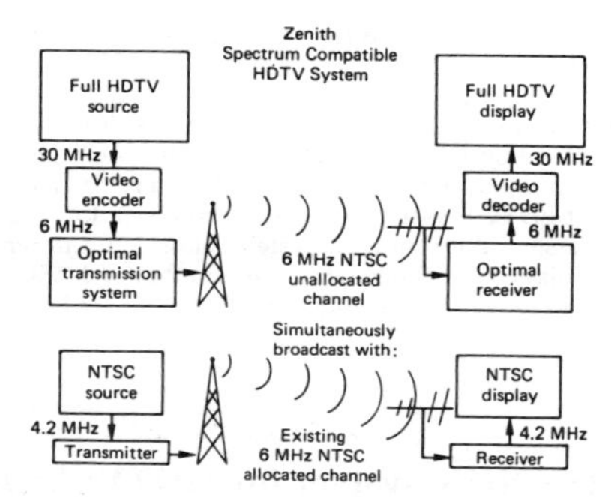

Figure 8.1 Zenith spectrum compatible HDTV system (Courtesy of IEE Publishing)

good reasons for this state of affairs. This is the way in which television broadcasting has evolved in the USA. As stated elsewhere in this book, the USA has a sophisticated network for the delivery of television, ranging from VHF/UHF to LPTV to MMDS, and finally to cable distribution networks feeding thousands of viewers from a single satellite master antenna system (SMATV). Moreover, viewers have the choice of anything from 30 to as many as 80 channels to choose from.

Given this background it becomes easier to understand that DBS will face a lot of competition. Terrestrial-based broadcasting will for a very long time be the mainstay of US television.

The United States Federal Communications Commission (FCC) is the governing body charged with the regulation of all broadcasting within the USA. In July 1987 the FC issued a Notice of Inquiry (NOI) concerning adoption of standards that would permit an advanced television service to be established in the USA. After studying responses the FCC in September 1988 issued a decision regarding a plan of action. This plan was further defined in 1990. In summary, the decision and plan of action specified that:

1. All existing terrestrial broadcasters must be able to implement ADTV.
2. The spectrum for such a service will be found in the existing television delivery services.
3. Advanced television broadcasting shall be able to be received by the present population of over 160 million NTSC receivers by using either: (a) an ATS signal that is compatible with and incorporates the NTSC signal within a single 6 MHz channel; or (b) an ATS signal that is accompanied by a simultaneous broadcast (simulcast) of an NTSC signal.
4. The channel bandwidth is limited to 6 MHz, leading to two possible plans. These are: (a) one existing 6 MHz single channel for an NTSC-compatible ATS signal; or (b) one new 6 MHz channel for an ATS signal plus an additional 6 MHz existing channel for the NTSC simulcast signal.

5. The FCC timescale for setting these new standards is summer 1994.

The consensus was that the first option will result in an enhanced-definition system (EDTV), while the two-channel simulcast plan would result in a true HDTV service. The advantage of EDTV is faster implementation with lower overall costs. Conversely, the simulcast system allows shut-down of NTSC transmissions at times when the number of NTSC receivers is no longer of a significant order.

Japan: the NHK system for HDTV: MUSE

NHK, the Japanese broadcasting authority, initiated research in the late 1970s. The result of this long research programme is the MUSE system of HDTV (Figure 8.2). MUSE, or Multiple Sub-Nyquist Sampling Encoding, was initially developed for use in DBS systems in the 12 GHz band. Since then MUSE has been expanded to be a hierarchical system which is suitable for terrestrial broadcasting in the UHF spectrum, CATV, VCR and video discs.

MUSE improves television picture quality and signal/noise ratio using techniques for motion compensation, quasi-constant luminance principles and non-linear emphasis along with band compression. The core of the different MUSE systems is MUSE T. MUSE T was developed for use with DBS, and needs a bandwidth of approximately 50 MHz for FM transmission.

MUSE in its original form was incompatible with any other television transmission standard to such an extent that no existing television receiver could receive MUSE or even be adapted. Thus the introduction of MUSE would generate a whole new market for HDTV receivers, in effect consigning existing television receivers to the scrap heap. As the Japanese electronics industry dominates the world's market in television receivers, the adoption of MUSE as a world standard for HDTV would have generated a massive market for domestic receivers which Japan would have dominated.

The NHK MUSE system is characterized by the parameters shown in Table 8.2.

When the NHK MUSE system was demonstrated to CBS in the USA the immediate reaction was that it provided true HDTV insofar as the picture quality was comparable to current high-quality 35 mm film. However, in the context of it being acceptable as the world standard there were a number of problems. The first was the very large bandwidth, in excess of

Table 8.2

Number of scanning lines	1125
Aspect ratio	5:3
Line interface ratio	2:1
Field repetition frequency	30 Hz
Luminance (Y) signal bandwidth	20 MHz
Colour difference signals bandwidth	7 MHz

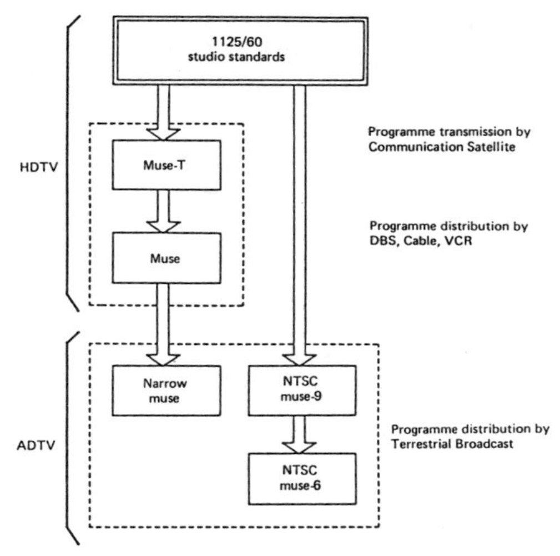

Figure 8.2 The structure of MUSE and its family (Courtesy of IEE Publishing)

30 MHz. The second was one of compatibility. Even if the required bandwidth was able to be compressed, the transmitted signal would be quite incompatible with prevailing standards. The third was the field repetition frequency. Whilst this would be suitable in North America, because of the 60 Hz electricity supply, it would present some problems in Europe because of the 50 Hz supply systems.

Since MUSE was first developed, and following some experimental demonstrations in Japan and North America, the problems mentioned earlier have been addressed. As a result, NHK has now introduced a MUSE family based on a hierarchical system using 1125/60 HDTV studio standard signals. Broadly, there are two divisions in picture quality, and three methods of transmission. As originally developed, MUSE was designed to be transmitted over a satellite link or by DBS. Since then the MUSE family has been extended to permit transmissions by cable, and by terrestrial transmission methods. With a view towards the North American television market, MUSE has been extended to permit ADTV transmissions and compatibility with the NTSC system in the USA (see Table 8.3).

MUSE-T is being developed for the transmission of television programmes between broadcasting stations. MUSE T employs a two-field cycle subsampling for motional resolution improvement. MUSE T has a 16.2 MHz signal bandwidth. NHK conducted some MUSE T transmission tests via the Canadian Anik C communications satellite in October 1987.

The ADTV systems are being studied for their improved picture resolution and for the extended 16:9 aspect ratio compared with the conventional NTSC system. NHK has developed two ADTV systems. One is NTSC-MUSE 6 and 9, which are compatible. The other is the non-compatible narrow muse. NHK has considered possible ways of expanding the aspect ratio for the ADTV systems by masking off the top and bottom to produce a 16:9 picture on the current 4:3 screen and by transmitting side pictures (Figure 8.3).

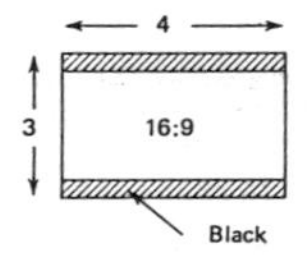

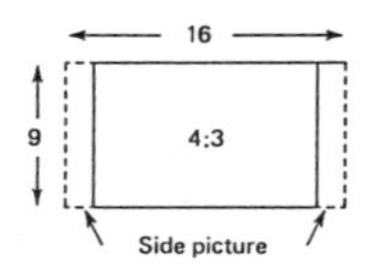

Figure 8.3 Expansion of aspect ratio for ADTV. (a) Top and bottom mask method. (b) Side picture method (Courtesy of IEE Publishing)

HDTV: the European approach

From the outset it was appreciated by the broadcasting authorities in Europe that HDTV must come by evolution and not by revolution. That is to say, it should come gradually in a number of technical phases so that not a single viewer is deprived of a satisfactory picture because of being unable to afford to invest in the latest technological advances. This is how it happened when colour broadcasting was first introduced into Europe. The PAL and SECAM transmission systems were so designed that viewers who could not afford to invest in a colour television set, which were very expensive when first introduced, were not deprived of a television transmission service, because the existing monochrome receivers could still receive a colour transmission, but it would be displayed in monochrome. As the television receivers gradually became cheaper through the benefits of mass production, so these viewers acquired colour televisions. Many economists would agree that the best way of building up a consumer market for a new product is by this method of evolution. Others would argue that the fastest method is to develop a product which makes existing appliances useless and therefore brings about a forced consumer market.

The European approach to HDTV is based on four separate phases over a period of a number of years so that eventually practically all of the viewers will have made the switch to HDTV. The basis of this approach was to develop a transmission system capable of HDTV and of being translated down to PAL and SECAM standards.

Phase 1 represents the low entry cost referred to above. The satellite dish and outdoor down-converter translate the signals at GHz frequency down to UHF and feed direct into a normal existing television receiver. This receiver will still be capable of receiving terrestrial transmissions at UHF frequencies from UHF-TV transmitters.

Table 8.3 *ADTV Systems*

	System	*Bandwidth (MHz)*	*Compatibility*	*Aspect ratio expansion*	*Resolution (lines/picture width)*	*HDTV/† ADTV aspect ratio*	*NTSC‡ aspect ratio*
HDTV	MUSE	9	No*	–	1020	16:9	16:9/4:3
ADTV	Narrow MUSE	6	No*	–	1010	16:9	16:9/4:3
	NTSC-MUSE-9	9	Yes	Top-bottom mask	900 ~ 960	16:9	16:9
				Side picture	900		4:3
	NTSC-MUSE-6	6	Yes	Top-bottom mask	680 ~ 960	16:9	16:9
				Side picture	900		4:3

* An experimental low-cost downconverter has already been realized.
† Aspect ratio displayed on HDTV or ADSTV receiver.
‡ Aspect ratio displayed on NTSC receiver.

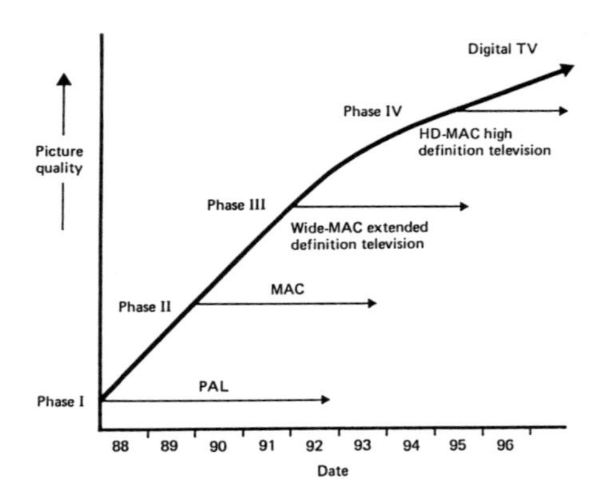

Figure 8.4 The phases in the evolution of MAC (Courtesy of IEE Publishing)

Table 8.4 *The European television market: viewers and systems*

Viewer population	*Percentage of total*	*Transmission*	*Standard*
130 million	82	Terrestrial	PAL/SECAM
25 million	16	Cable delivery	PAL/SECAM
2.5 million	1.5	Medium power satellite	PAL/SECAM
150 000	0.5	DBS satellite	MAC

World standard for the introduction of HDTV

Ten years ago it seemed possible that when HDTV was introduced the world's broadcasters and television manufacturers would have seized the opportunity of deciding on a single HDTV standard. That was when Japan was developing its MUSE HDTV. However, with the formation of the Eureka project and D-MAC the prospects of a single world standard diminished.

The decision of the USA to develop its own HDTV standard has finally killed any prospect of a single world standard ever coming about. This will mean that, as with the existing NTSC/PAL/SECAM standards, the world's television manufacturers will be producing HDTV receivers to satisfy the three different markets: Japan and the Far East, Europe and America. So, is there merit in the USA going ahead with its own HDTV? From a technology standpoint the answer must be yes. Both the Japanese MUSE and the D-MAC system are analogue-based whereas any future system in the USA will take full advantage of digital techniques.

In effect this will mean that the USA will have leapfrogged Japan with its almost 15-year-old technology. The future of

Table 8.5 *Distribution systems for HDTV on a worldwide basis*

HDTV and Region	*Cable*	*Satellite*	*Terrestrial*	*VCR*
MAC Europe	Yes	Yes	No	No
MUSE Japan	Yes	Yes	No	No
Digital USA	Yes	Yes	Yes	No

television broadcasting is certainly going to be exciting and with a few surprises along the way. What is a certainty is that the standards manufacturers will still be in business. Manufacturers are gearing themselves for future needs.

PALplus: precursor or competitor to the MAC format?

PALplus was conceived in 1989 as a precursor to HDTV in Europe. As the name implies, it is an enhancer of the existing PAL television standard. The strategy is quite sound and is the brainchild of the PALplus Steering Committee, which is drawn from broadcasting authorities and major manufacturers.

The PAL system came into use in 1967 and has served Europe well; however, since the advent of satellite television and, more recently, DBS broadcasting, these transmission systems are, by virtue of the wider bandwidth, able to support the transmission of HDTV systems such as D-MAC and D2-MAC. For one reason or another there has been a reluctance by broadcasting authorities to make the switch from broadcasting programmes in PAL to one of the two MAC systems. HDTV will come. Its most enthusiastic supporters are the manufacturers of television receivers, who quite naturally are keen for HDTV to be introduced into Europe. To this end, three of the four major European manufacturers have already begun production of HDTV receivers with a 16:9 screen aspect ratio; they are Grundig, Philips and Thomson. These television receivers represent a new generation providing image and sound quality which have to be experienced in order to be able to appreciate the giant stride that has been made in receiver design. Additionally, all will be able to present pictures enhanced in 16:9 image format corresponding to the live cinema experience. A major factor in this optometric and physiological enhancement is the large screen.

The public broadcasting authorities of Switzerland, Austria and Germany formed the PAL strategy group in order to take advantage of these developments in television broadcasting. The broadcasting authorities are ARD, ZDF, ORF, SRG and IRT. These broadcasting authorities are working in conjunction with Grundig, Nokia, Philips and Thomson. In September 1989 the PAL strategy group agreed on a working programme to react to the developments in HDTV outlined earlier. One of the principal aims of PALplus has as its objective the adapting of terrestrial-based television broadcasting in the VHF/UHF spectrum to be able to transmit colour television programmes that in both quality of picture and wider image format correspond more closely with those high-definition

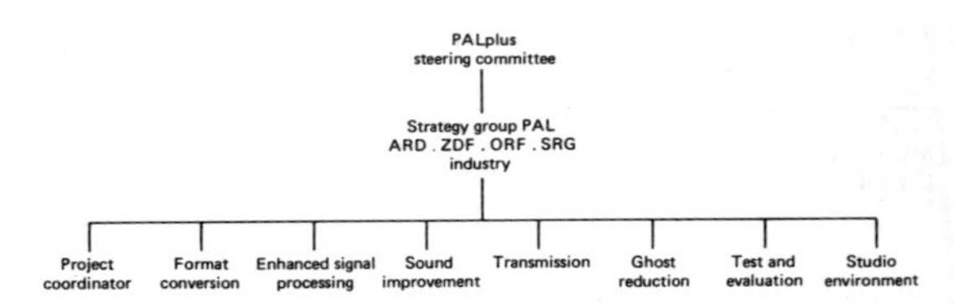

Figure 8.5 Structure of the work (Courtesy of Palplus)

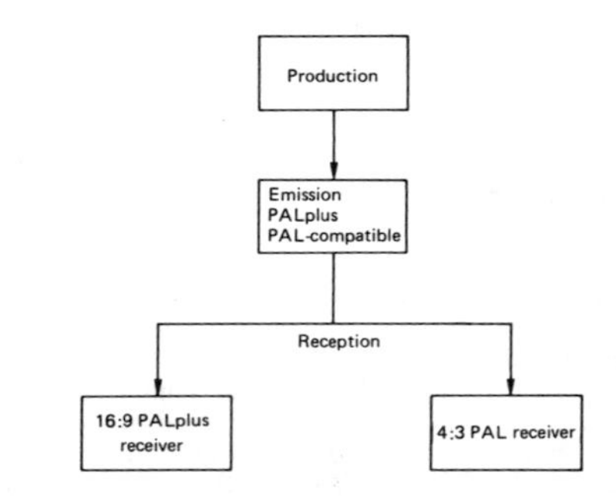

Figure 8.6 Transmission and reproduction processes of PALplus signals (Courtesy of Palplus)

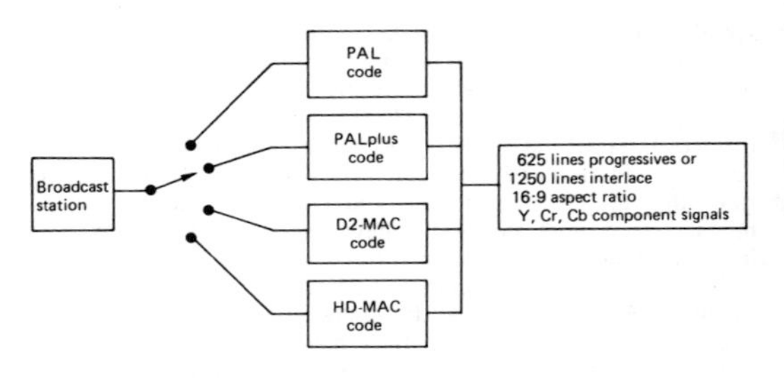

Figure 8.7 Structure of the future studio (Courtesy of Palplus)

programmes which will be available from satellite DBS transmissions.

In essence the PALplus system is intended to marry the 16:9 aspect ratio to the existing PAL format of 4:3. Additionally, the PALplus system will improve picture resolution and quality of sound reproduction, whilst at the same time eliminating the technical weaknesses in PAL, i.e. the effects of cross-colour and cross-luminance artifacts which detract from an otherwise acceptable picture presentation.

Figures 8.5 to 8.8 show respectively the organization, the structure and the technical programme, the transmission and

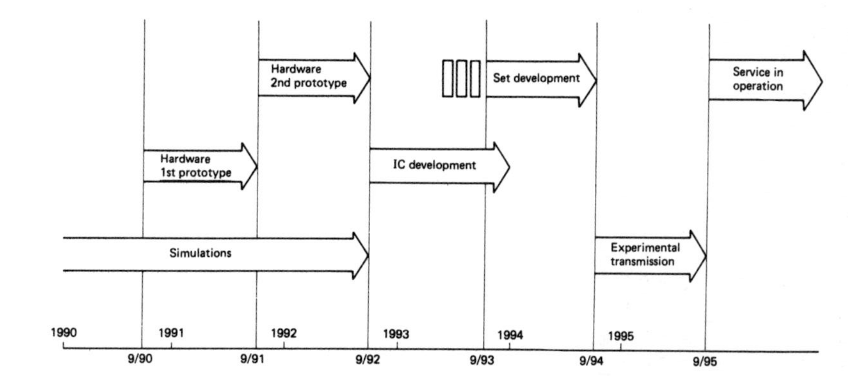

Figure 8.8 PALplus introduction (Courtesy of Palplus)

reproduction process, the studio production, and the hardware development programme from 1990 to 1995.

The development programme for PALplus is designed in such a way that any steps for improvement can be pursued and introduced independently of one another. Top priority in the process is full compatibility with existing PAL transmissions from terrestrial-based transmitters. This means that PALplus must be capable of being transmitted in the existing 7 MHz channels used in VHF/UHF television bands III, IV and V, and received on existing television sets. The most important step, however, is that PALplus will show a wide screen 16:9 aspect on existing television receivers. This will be achieved by sophisticated software which will blank off the top and bottom of the television screen. This technique is known as the 'letter box' display. Such techniques have their advocates and critics, but it is generally thought that by the year 2000 the market penetration by 16:9 television receivers will be about 40%, rising to a figure of 88% by the year 2005. On these figures the adoption of PALplus over terrestrial networks would seem to be fully justified.

9 Direct broadcasting by satellite (DBS)

System considerations

Any transmission system is a compromise between a number of different factors. These include:

(a) the frequency in use
(b) the bandwidth available
(c) the power available in the transmitter
(d) the desired quality of service to the consumer

In satellite broadcasting of television signals there are two types of transmission service which have achieved prominence in the past decade:

(a) TVRO and SMATV systems
(b) direct broadcast by satellite to the homes of viewers

The fundamental difference between the two systems is the design of the receiving terminal. Television receive only (TVRO) and satellite master antenna systems are intended to receive television signals and then relay these to the homes of viewers, usually by cable systems. Because there is only one television receiving system, it can be professionally designed and engineered, and can also use very large dishes compared to domestic viewers. These dishes, using Ku-band frequencies, can be 3–4 m in diameter. Similarly, the television receiving system can be made much more sensitive than a domestic television receiving system costing up to 500 dollars.

In the context of DBS, things are very different. Television receiving systems must be costed within the reach of the average viewer. The television satellite dish must be small, 50–90 cm in diameter, unobtrusive and cheap to erect and install.

All these factors combined mean that a domestic satellite installation cannot compete on equal terms with TVRO/SMATV systems. In order to compensate for this inequality between professional and domestic-type satellite receiving systems, WARC 77 laid down certain standards for DBS service, one of these being the necessity for a greater output from the transponder.

For a DBS service the broadcaster can employ transponders with output powers up to 240 W. Such an output power would guarantee a very good service to the viewer. It would, however, impose very severe constraints on the design of the satellite, and it would, as a result, increase very considerably the amount of investment by the broadcasting authority or company and the programme providers.

At the other end of the scale the broadcasting authorities can elect to use as low a transponder output power as is possible to meet a certain grade of service for 99% of the time.

This permits a greater number of transponders to be operational at any one time.

Direct broadcasting by satellite in Japan

Japan was the first nation to start up regular television broadcasting through DBS direct to the homes of viewers. Though this achievement might have seemed an obvious thing to many, because of Japan's leadership in electronics, communications and studio production technology, there are other reasons, and these are to do with the topography of Japan.

Japan is composed of many islands, both large and small. Its geography is harsh. Outside the principal cities it is characterized by valleys, mountains, rivers and very steep ridges. These features make it very difficult to set about planning an efficient terrestrial broadcasting system for television. Whether by VHF or UHF, the problems remain the same: how to reach the people. Both VHF and UHF require good line of sight, a commodity in short supply in Japan. Although a very efficient network of terrestrial-based VHF/UHF-TV transmitters serve the people, the cost has been very high and the technical problems enormous. Even then it has not always been possible to eliminate problems associated with line-of-sight transmissions, including that of signal ghosting.

Cable television

Given the problems of operating and maintaining a reliable terrestrial broadcasting network it was a natural consequence for Japan to explore the advantages of delivering television signals to viewers by cable. One of the main reasons for its introduction was to solve the problem of ghosting.

By 1988 the total number of cable television systems was 40 000, reaching 4.5 million viewers. A few of the cable networks were large-scale systems with a capacity to reach out to as many as 10 000 viewers. Even with this very powerful cable network capacity, and the previously existing VHF/UHF-TV networks operated by NHK, there was still a shortfall, estimated at between 420 000 and 1 040 000 households left without a satisfactory service. These figures exclude those households with a television service but who still experienced ghosting problems. The two islands of Ogasawara and Daito, both located in the Pacific some 1000 km from the nearest main islands, had neither a VHF/UHF or cable television facility.

Development of a national satellite delivery system for television

From the foregoing overview of Japanese television, it becomes easy to see that there was a clear incentive to develop a national satellite delivery system. There was also another reason; a satellite delivery system was a necessary precursor to an HDTV service. It was also healthy for the industry itself to develop alternative systems. Bringing television to those millions who previously had no service whatsoever could only generate a healthy receiver-manufacturing industry, thereby lowering television receiver costs, and promoting the development of newer models suitable for the introduction of HDTV.

The BS-2 direct broadcast satellite Following the recommendations of WARC 77 meetings, conceived for the express purpose of defining the regulations that would apply to satellite broadcasting, Japan was the first country in the world to take appropriate action. WARC 77 laid down the following plan for the three main regions of the world, these being:

Region 1. Europe, Middle East, Africa, CIS (then USSR) and East Asia.
Region 2. North and South America and Greenland.
Region 3. India, Iran, South East Asia, Australasia, Japan and China.

Region 3, in which Japan is situated, was allocated the satellite frequency band of 11.7–12.2 GHz.

Operational satellite broadcasting began in Japan in 1980 in full accordance with the wishes and hopes of WARC-77. The Japanese Space Agency NASDC launched the satellites from its Space Center at Tanegashima by N-11 launch vehicles. Satellite BS-2a was launched successfully in January 1984 and the second satellite in the series was launched in February 1986. The overall satellite project includes satellites, tracking, telemetry and control station (TTC), a number of earth stations TVROs and terrestrial re-transmitting stations, and community receiving stations.

By the use of specially designed antennas on board the satellites it became possible to cover the whole of Japan with a single satellite. The power output from the transponder is 100 W and the EIRP is of the order of 58 dBW.

System performance Within the primary reception zone, which includes the heartland of Japan, good reception was obtained with dishes of 50 cm diameter. In the intermediate reception zone satisfactory reception is possible with 75 cm dishes. Only at the extremities of the Japanese islands was it necessary to employ 90–120 cm dishes. This region is noted for thunderstorms, typhoons and heavy rainfall, and when these occur there is some deterioration in quality of signal reception, but under average conditions of rainfall there is an adequate margin in system performance. NHK started its satellite broadcasting in May 1984, transmitting 18 hours per day and has since expanded to two-channel broadcasting by using satellite BS-2b. By so doing NHK was successful in promoting the use of satellites to reach those sectors of Japanese viewing audiences which until then had been deprived of a satisfactory television service.

The history of radio broadcasting has shown that developments in the technology of transmission, making it easier for listeners to receive distant signals, contributed to the proliferation of receivers with a better performance. There is no reason to suppose that satellite broadcasting of television programmes will not have the same effect.

Satellite broadcasting of HDTV

NHK has been actively engaged in the development of HDTV since the early 1980s. To begin with, the early transmissions were over a closed circuit, but from 1986 NHK felt confident that it could move towards the use of satellites. The Japanese satellite BS-2b was used for these experiments. The transmitted signals using the MUSE system were received at various places all over Japan and confirmed that MUSE, like the NTSC transmissions that had gone before, could be received equally satisfactorily. Between 1986 and 1988 further proving trials took place. These tests included test transmissions over an Intelsat satellite, the Anik C satellite of Canada, and the RCA K1 satellite. Satisfactory reception was achieved at Ottawa, Montreal and Toronto in Canada, whilst the RCA satellite's footprint covered cities as far apart as New York, Washington, Los Angeles, Seattle and Victoria Island (see Figure 9.1 and Table 9.1).

With a dedication to DBS unmatched by any other nation, the Japanese broadcasting authority NHK in conjunction with NSDA, the National Space Development Agency, launched the first in the BS-3 series of satellites, and this was followed in 1991 by a second BS-3 satellite. Both these satellites were built by General Electric for NHK and launched from the Japanese Space Facility in the southern prefecture of Kagoshima.

NHK expects a healthy growth rate in viewers investing in receivers for reception of DBS broadcasts. The history of broadcasting shows that easy and inexpensive reception greatly contributes to a proliferation of receivers, with improvements in technology going hand in hand with lower prices for the consumer. This is the policy of Japanese broadcasting and is expected to result in a market of three million viewers of DBS by the early 1990s.

Japanese experiments in broadcasting of HDTV by direct broadcast satellite The Japanese broadcasting satellite BS-2 was used for the experiments. Experimental equipment, including both

Table 9.1 *MUSE experiments*

Period	*Content*
Dec. 1986	BS-2b
Jan. 1987	Washington DC, USA Am-VSB FM
Feb. 1987	Two-point reception, BS-2b
March 1987	BS-2b daytime experiment
May–June 1987	Hi-Vision Fair, BS-2b 22-GHz bandwidth FM transmissions FM-CATV
Aug. 1987	Intelsat transmission
Aug. 1987	Yume Kojo (Intelsat)
Oct. 1987	Canada/USA CS two-hop transmission and CS-FM-CATV
Nov. 1987	Hi-Vision Week BS-2b including the transmission from Osaka (daytime)
Dec. 1987	BTA experiment (BS-2b) Reception by several companies (night-time)
Feb. 12–24, 1988	CS experiment

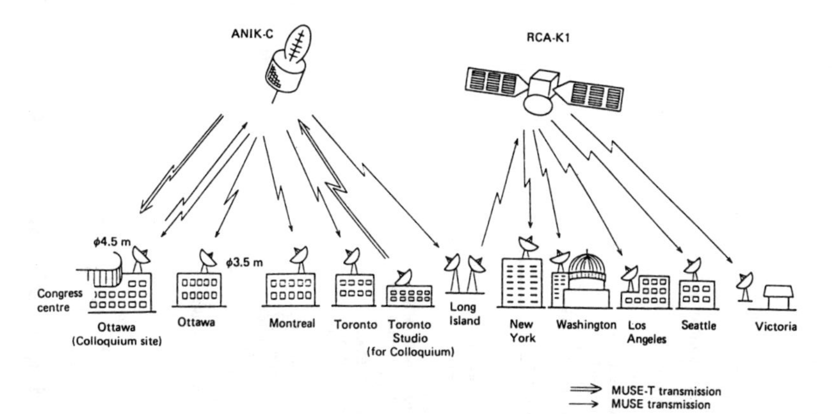

Figure 9.1 Outline of transmission experiments (Courtesy of IEE Publishing)

transmitter and receiver, was installed at the NHK Broadcasting Centre in Tokyo. The programme source, an HDTV signal, was band compressed to a baseband width of 8.1 MHz. Table 9.2 shows the calculated link parameters for the reception in Tokyo under the conditions of good weather and clear visibility. The receiving system was based on a receiver dish size of 0.75 m with an efficiency of 65%, a receiver noise of temperature 260 K and *G/T* of 13 dB/K.

Table 9.3 shows the link budget at the coverage edge (−3 dB from the centre gain of the transmitting antenna).

Table 9.2 *Calculated link parameters for the experiments*

Frequency (GHz)	12
Type of modulation	FM
Equivalent RF bandwidth (MHz)	27
TWT output power (W)	100
Satellite antenna gain (dB)	39
EIRP (dBW)	56.7
Rain attenuation (dB)	0
Atmospheric absorption (dB)	0.1
Free space attenuation (dB)	205.6
Feeder link contribution (dB)	0.3
Receiving equipment *G/T* (dB/K)	13
C/N (dB)	18
S/N unweighted (dB)	39.7

Table 9.3 *Link budget of satellite transmission in MUSE system*

Frequency (GHz)	12
Type of modulation	FM
Equivalent RF bandwidth (MHz)	27
C/N (99.9% of worst month) (dB)	17
S/N unweighted (dB)	39
Figure of merit *G/T* (dB/K)	16
Required power flux density (dBW)	−110.5
Free space attenuation (dB)	205.6
Rain attenuation (dB)	2
Atmospheric absorption (dB)	0.1
Feeder link contribution (dB)	0.3
Coverage edge area factor (dB)	3.0
Required EIRP from satellite (dBW)	57.7
Satellite antenna beamwidth −3 dB (dBi)	1.3 × 1.8 degrees
Satellite antenna gain (dBi)	40.0
Losses, feeders, filters, etc. (dB)	2.3
Required TWT power (dBW)	20.0
Watts (W)	100

Broadcasting by DBS in the USA

The USA has a mature television distribution system, composed of high-power VHF/UHF, low-power television (LPTV), multipoint, microwave distribution systems (MMDS), super-power UHF-TV stations, and finally the cable distribution networks which are also expanding rapidly, to a point where in a few years time the penetration figure will be 75% (Table 9.4).

Against such a vast choice of delivery systems for television programmes – and it must be remembered that Americans generally have a choice of 30 or even in some cases 80 channels of viewing when connected to cable systems – any investment into yet another source of programme delivery must inevitably be seen as a high-cost and a high-risk investment. This makes it difficult to find financial backers.

However, things are moving ahead with proposals for the introduction of DBS and a number of commercial operators have made proposals in this direction.

The USA has a history of 70 years in radio and television broadcasting. Its structure has emerged through a policy of successive governments of exercising the absolute minimum of state control. In other countries such a policy might have been a recipe for chaos, but not so in the USA. A combination of geography, history, economics, politics, and not least the sheer size of the country, has resulted in a domestic broadcasting network which is in fact composed of a number of broadcasting networks commercially owned and operated.

The USA has over 11 000 radio stations and over 1400 television stations, most of which are affiliated to one of five major networks. Herein lies the key to the success and the high technical standards of American broadcasting. From the very beginning of American broadcasting, an industry evolved to suit those needs.

From 1948 to the mid-1980s the broadcasting networks of NBC, CBS and later ABC exercised almost total control of live interconnected commercial television programming within the USA. In addition to the 14 000 television stations there are over 400 LPTV stations operating.

The USA is very large, stretching about 4500 km from east to west, and 2575 km from north to south with 48 contiguous states. It is often not realized outside the USA that there are quite strong differences in character and culture between the peoples of these 48 states.

In any broadcasting network it was essential not to ignore local character. The radio networks conceived by NBC and AT&T permitted this by enabling any radio station to broadcast local news and locally produced programmes when not taking the main programme feeds from NBC's studios in New York.

When the television age came to the USA, developments in network programming and switching systems were given further impetus. Telephone cables gave way to a vast coaxial

Figure 9.2 Typical broadcast installation in the USA (Courtesy of Andrew Corporation)

cable system and after UHF television broadcasting was introduced the networking authorities changed to microwave relay systems, which were able to provide a much better quality of service.

The USA also has a large and growing, privately owned cable television industry. It is made up of over 5000 local cable systems which pass on 50% of all US programmes to these subscribers, and it is confidently expected that this figure could rise to as much as 80% within the next decade. Many of these cable networks receive their television programmes through master antenna satellite systems. There are currently over 160 million hómes in mainland USA representing a total investment of the order of 1 billion US dollars.

This huge sum represents the initial cost of nearly 200 million colour television sets, a number which is increasing at the rate of 20 million each year. Between now and the year 2000 over 300 million more television sets and VCRs are the projected requirement. Many homes in the USA have more than one television set, but all have one feature in common: they are all to NTSC standard, i.e. 525/60 Hz.

More so than any country, the privately owned television stations compete with each other and against other forms of media for the attention of viewers. In order for that competition to be effective they must deliver on two counts. These are (a) programme quality, and (b) picture quality.

Requirements for HDTV in the USA

Any proposal for HDTV, or advanced television (ATV) as it has become called in the USA, must take account of the interests of all parties. These are:

(a) the programming networks such as NBC, CBS and others
(b) the networking authorities
(c) the broadcast and television station owners and managers
(d) the advertisers and sponsors
(e) the consumer

The final solutions must make economic sense for all, which means that, whatever form of higher definition system is employed, it has to be based on NTSC, that is to say existing NTSC televisions must not have to be consigned to the scrap heap.

The principal and ultimate force behind the implementation of any form of ATV must be picture quality.

Pictures on screen look terrific at 10 inch, and very good at 21 inch, but if a three-foot screen were to be used the picture would look pretty poor. As the picture size goes up, the information density goes down, and the overall picture quality

Table 9.4 *Statistics as at 1989 for radio, television and cable television in the USA*

Television stations	VHF 545	UHF 508	Educational 335	Total 1388
LPTV stations	VHF 126	UHF 329		Total 455
Radio stations	AM 4929	FM 4141	Educational 1369	Total 10 439
Cable television	Subscribers 48 637 000	Penetration 53.8%	Pay penetration 32%	

Source: Broadcast Engineering, May 1989

diminishes, leaving both eye and brain less than satisfied. Another important quality is the brightness of the picture; a three-foot screen is not going to be any good if the brightness is less than that of a 10-inch screen.

Finally, there is the overall cost of a new ATV television set to consider. Experts do not agree on what the ceiling price should be, but all are agreed that within a few years after the introduction of ATV the prices of receivers should fall as the demand for sets increases.

The status of television in the USA

Three-quarters of the 1400 television stations are commercial. Ownership is highly diversified; the FCC has restricted station ownership by any one person or company to a maximum of 12 stations or coverage of 25% of the nation's population. The four major networks, NBC, CBS, ABC and Fox, are affiliated with over 800 of the total number of television stations. The remaining television stations, about 600, are classified as 'independents'; that is to say they broadcast mixtures of recorded local and special event programming. However, the tendency is for more and more television stations to become affiliated to one of the four networkers.

Nearly 60% and rising of the population have VCRs. Of those who receive television from cable systems (about 50%), 75% can have the option of watching any one of 22–35 channels with less than 10% of viewers having more than 35 channels. However, as regards DBS, less than 3% of US citizens have this capability.

Introduction of DBS in the USA

DBS is developing fast, and the position with regard to fibre optic systems is no different. A survey of communications in the USA shows that there is every possibility of the major cities being connected by fibre optic long-haul communications within the next ten years. The growth of domestic fibre optic systems will follow on rapidly. For DBS to succeed against this scenario it must offer the same advantages as fibre, which means a wide choice of viewing channels.

The SkyCable consortium has recently announced plans for the launching of a high-power satellite capable of providing as many as 108 NTSC channels and including capacity for HDTV services. The launch date is set for 1993, with programming services to begin in 1994.

The launch of the United States' first high-power direct-to-home satellite was assured by the signing of a contract between Hughes and Arianespace calling for an Ariane rocket to lift Hughes' first DirecTv satellite into orbit in December 1993.

The advanced Hughes satellite will deliver news, sports, movies and specialty television programming directly to households equipped with low cost, 18 inch receive antennas. Located at the 101°W longitude orbital position, the DirecTv satellite will provide a national broadcasting platform for 48 states.

With DirecTv, programmers will have an opportunity for direct access to 92 million households in the United States.

The launch contract follows a previous agreement between Hughes Communications and United States Satellite

Broadcasting, Inc. (USSB). In June 1991, USSB invested more than $100 million and purchased a five transponder payload on Hughes' first DirecTv satellite. The agreement also provided USSB with access to the digital compression and encryption technologies to be used for the Hughes' DirecTv system.

Hughes Communications holds an FCC licence to operate 27 direct broadcast frequencies from the 101°W longitude orbital location. USSB holds a permit for the remaining five frequencies at 101°. Through the use of digital compression, multiple channels of video programming can be transmitted through a single direct broadcast frequency.

Two DirecTv satellites for Hughes Communications are currently under construction by Hughes Space and Communications Group. Based upon Hughes' HS 601 design, each satellite will carry 16 operational transponders and feature 120 watt transponders broadcasting in the DBS portion of the Ku-band frequency. Upon launch of the second satellite, more than 100 channels of television programming will be provided from the 101°W longitude orbital location.

With each passing month the DBS momentum gathers more support from broadcasters. Improvements in satellite technology coupled with the prospect of digital video compression are the spur to the interest by broadcasters and the programme distributors.

The advent of the high power DBS satellite coupled with another stunning technology, digital video compression, brought broadcasters back into the arena. As of today no less than eight applications for high power DBS systems have been submitted to the FCC in Washington DC (Table 9.5). These give a total of 20 satellites, all with high power, 16 transponder channels.

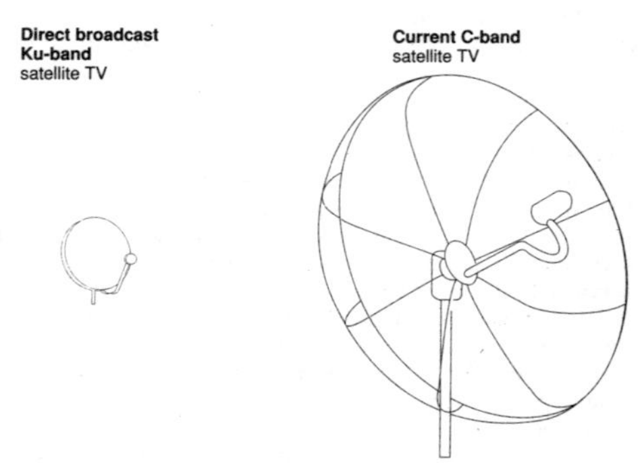

Figure 9.3 Comparison of DirecTv Ku-band and C-band configurations (*Source:* L. Lockwood, DirecTv, a digital DBS, *International Cable*, April 1993)

Table 9.5

Company or consortium	*Proposal and quantity*	*Date*
Loral Corporation	3 high power DBS sats	
Hughes Communications	2 high power DBS sats type HS-601s	Orbital slot 101°W launch date December 1993
Echostar Corporation	7 Series 7000 GE Astro high power DBS sats	Orbital slot 119°W launch date 1995
Advanced Communications Corp and Echostar	2 Series 7000 GE Astro	Orbital slot 119°W launch date 1994
SSE Telecom	2 high power DBS sats	Scheduled for 1996
Direct Broadcast Corporation	2 high power DBS sats	Scheduled for 1996
Dominion Video Satellite	1 high power DBS sat	Planned for 1994
Temp	1 high power DBS sat	Planned for 1995

HDTV technology: analogue versus digital?

Is it conceivable that the USA, thought by many to be unable to match the technology from Japan and Europe, will in the very near future be able to outstrip both as regards HDTV technology? Could it be that Japan's NHK, one of the leading broadcasters in the world in regard to broadcasting technology, is about to see its 20-year lead in HDTV be superseded by American technology. Could it be that Europe's insistence on D-MAC and D2-MAC was a wrong decision? And finally, could it be that the rest of the world will adopt the final selection of the FCC when it gives its ruling on which of the six HDTV systems will be adopted as the US standard?

All of these questions are waiting to be answered as the race to develop HDTV goes ahead. From being a non-starter, the USA has jumped to the front of the starting grid.

Technology is making such fast progress in the field of HDTV that it becomes necessary to pause and think on what kind of delivery systems the societies of tomorrow will demand. Ten years ago the satellite 'bird' was the ultimate, largely because of its ability to carry wideband transmissions that HDTV demanded; 27–40 MHz, well beyond the capability of any other transmission medium. Then along came the prospect of compressing HDTV signals down to smaller bandwidths.

Given the popularity of cable delivery systems in many European countries, and particularly in the USA it seems that what is needed is a delivery system that can work satisfactorily with satellite, terrestrial-based television transmitters, i.e. UHF-TV, and also cable systems, coaxial, optical fibres or a combination of the two. By a happy coincidence the five systems under evaluation by the FCC will meet all three requirements.

10 Multiplexed analogue component systems (MAC)

MAC systems

Though it is now a virtual certainty that Europe will take the digital route to HDTV in the future, it is also the case that D-MAC will be in use until beyond the year 2000. For this reason this chapter on the MAC variants is included. At the present time there are 11 users of the MAC systems.

The evolution of the MAC family of television transmission standards began with the development work carried out by the Independent Broadcasting Authority (IBA) in the UK, in conjunction with other European broadcasting authorities, and much of the very considerable research effort was funded by the European Broadcasting Union (EBU). This research and development programme resulted in a range of standards, each differing slightly in detail, but essentially similar in overall concept. The MAC/packet family was introduced to meet the requirements laid down by the WARC conference for a system suitable for transmission by direct broadcast satellites and cable.

The MAC/packet family consists of the C-MAC system, the D-MAC system and the D2-MAC system. For all these the basic principles of operation are as follows. The basic colour television system produces a picture by the use of a scanning spot, which produces an image by scanning across the television screen. The quality of this image is the sum total of three parameters:

(a) the number of lines per picture
(b) the picture presentation rate
(c) the aspect ratio of the picture

An increase in any one, but preferably all three, will result in an enhanced picture quality. The colour picture is made up of two basic elements: luminance and chrominance. Luminance controls the range of brightness, whilst chrominance controls the fidelity of the picture in terms of correct colour presentation.

One of the reasons for the development of a new family of television standards was to eliminate defects and deficiencies in the existing PAL, SECAM and NTSC standards which have been in service since the mid-1960s. One of the worst failings of these existing systems is due to the fact that the luminance data and chrominance data were sent together. The effect of this was cross-modulation, with one set of data affecting the presentation of the other. This results in the well-known effect of dappling and colours running into each other (see Figure 10.2).

To overcome these serious defects the MAC system transmits three pieces of data: the sound and sync data, the colour difference data, and the luminance data. These are transmitted as a data stream lasting 64 μs. (see Figure 10.3).

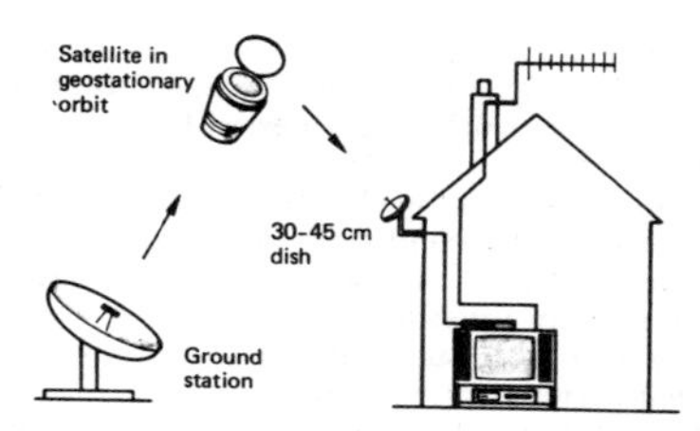

Figure 10.1 DBS reception in the home. The high-power transmissions from the UK satellite will allow reception on small dishes. The indoor receiver unit includes the D-MAC/packet decoder, and can be used with an existing television set

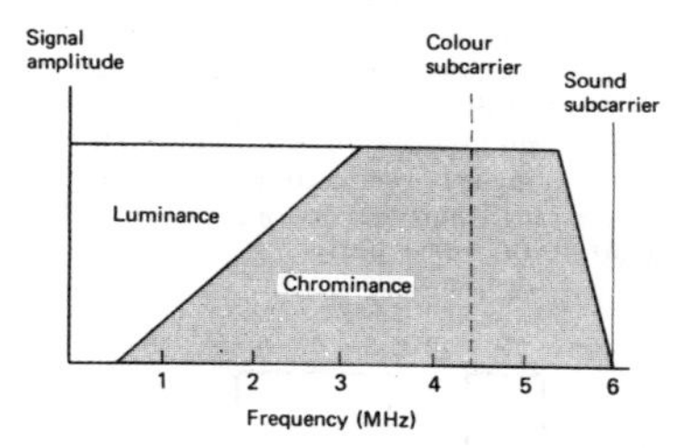

Figure 10.2 Baseband spectrum of PAL system I. Adding a colour subcarrier introduces spurious patterning effects. Also in FM systems areas of saturated colour are subjected to a disproportionate amount of noise (Courtesy of IBA)

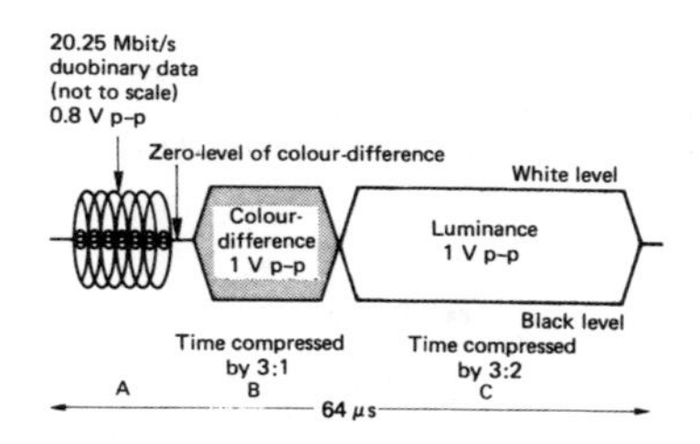

Figure 10.3 Baseband video waveform of a D-MAC television line. The duobinary data burst conveys 206 bits and is followed by time-compressed vision signals (Courtesy of IBA)

Section A is a data burst that carries both sound and sync information; this data lasts for 10 μs and has 206 bits per line, giving an instantaneous data rate of 20.25 MHz. This signal will carry eight high-quality sound channels.

Section B contains the colour difference components, with alternate lines containing U (blue difference) and V (red difference) components. The signal is processed so that each colour difference line is compressed in time by a factor of 3:1 thus occupying a time-span of 17.5 μs. Section C contains the luminance information which is compressed by a factor of 1.5:1 to occupy a time period of 35 μs.

This description relates to the C-MAC system designed to be transmitted over a 27 MHz bandwidth in accordance with WARC 77 recommendations. The advantages of C-MAC are: immunity from non-linearities associated with FM, zero cross-luminance distortion, suitability for HDTV, and compatibility with scrambling systems.

The B-MAC is another version. It is formatted as in the C-MAC system but with a difference in the data burst that carries audio and sync information. B-MAC can carry six high-quality digital audio channels, teletext, messages, and control services for service and subscriber information.

The basic principle behind the MAC system is the use in time sequence of separate signals for luminance and colour difference. There are no subcarriers; instead the luminance and colour difference video signals are time-compressed before being transmitted. Within the time occupied by one conventional line period (64 μs for a 625-line system) it becomes possible to include one colour-difference component. Alternate lines contain U and V (the blue and red difference signals) followed by the luminance.

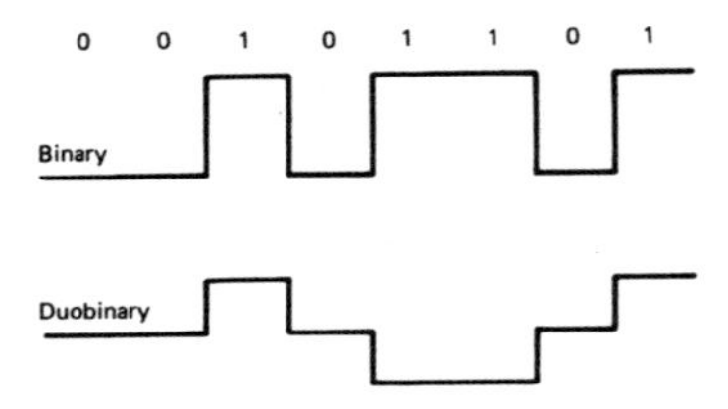

Figure 10.4 Comparison of binary and duobinary coding. After low-pass filtering, the three-level duobinary data signal is time-division multiplexed with the time-compressed vision signal (Courtesy of IBA)

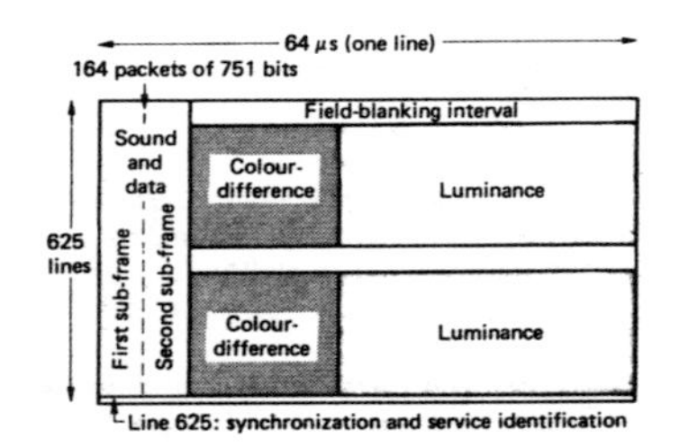

Figure 10.5 Simplified structure of D-MAC/packet frame showing the distribution of vision and 20.25 Mbit/s duobinary data. Organizing the data into two sub-frames allows easy transcoding of one sub-frame into D2-MAC (Courtesy of IBA)

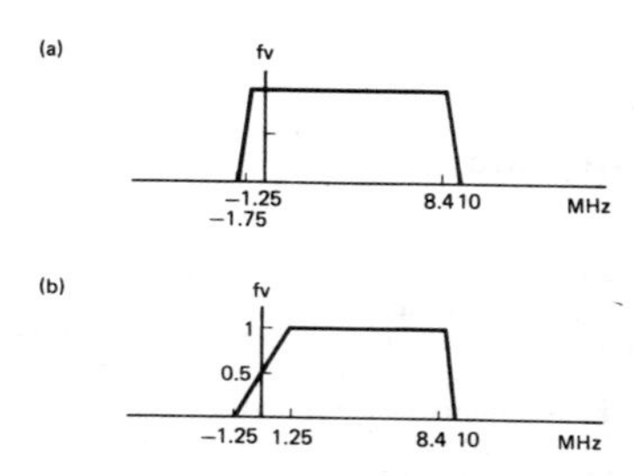

Figure 10.6 D-MAC on cable: (a) proposed response of an ideal transmitting filter for a 12-MHz channel and (b) receive filter provides full VSB shaping (Courtesy of IBA)

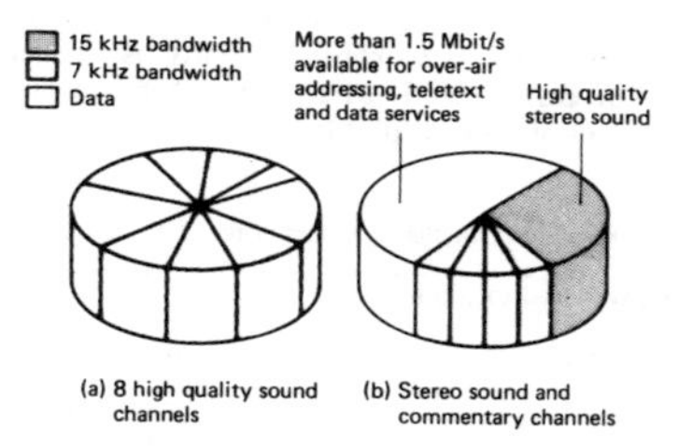

Figure 10.7 The high data capacity of D-MAC/packet (approximately 3 Mbits) allows great flexibility. Two of the many possible options are illustrated (Courtesy of IBA)

Each line of luminance is compressed by a factor of 3:2 and this reduces the time occupied from 52 to 35 μs. Similarly each line of colour difference is compressed by a factor of about 3:1, reducing the time to about 17.5 μs.

This process of time compression results in an increase in baseband bandwidth. For luminance the maximum uncompressed video bandwidth is about 5.7 MHz. Following compression (prior to modulation) the picture signal occupies a bandwidth of about 8.5 MHz.

Unused portions of the vision signal before the start of each picture line carry a high-capacity digital data signal in place of the wasteful sync pulses of conventional television signals such as in the PAL system. The data signal occupies about 10 μs of each line period and consists of a 20.25 Mbit/s duobinary data burst inserted at baseband. Each line carries 206 bits of data, resulting in a high mean data capacity of about 3 Mbit/s. Line 625 is occupied entirely by the data signal and conveys 1296 Mbit/s.

High-quality sound

The digital sound system which is a feature of MAC offers high-quality stereo with the possibility of additional multi-language sound channels. High-quality sound with an audio bandwidth of 15 kHz (using a digital sampling rate of 32 kHz) is offered by either linear coding of 14 bits per sample, or near-instantaneous companding (NICAM) of 14 bits per sample to 10 bits per sample. The latter option is the coding method that has been adopted for digital stereo on the UK terrestrial UHF-TV network and is the preferred option for DBS in the UK.

Medium-quality sound can be used to provide commentary sound channels where high-quality sound is not mandatory. The audio bandwidth in this case is 7 kHz and the digital sampling rate is reduced to 16 kHz. As many as 16 sound channels could be carried by D-MAC if the data multiplex is related to this level of sound quality.

Table 10.1 *Summary of the main characteristics of the D-MAC format*

Carrier/noise ratio required to produce a grade 4 picture	12 dB
Number of lines per frame	625
Interlace	2:1
Horizontal frequency	15 625 kHz
Vertical frequency	50 Hz
Synchronization	Digital, in data burst
Luminance bandwidth (uncompressed)	5.75 MHz
Chrominance bandwidth (uncompressed)	2.75 MHz
Chrominance compression	3:1
Aspect ratios, choice of two:	4:3 or 16:9 full screen
Sampling frequency	20.25 MHz
Rate of digital burst	20.25 Mbit
Total data capacity	3.08 Mbit
Maximum number of high-quality sound channels	8

Table 10.2 *The main advantages of D-MAC*

Vision, sound and data exist as a single wire baseband signal
Considerable improvement in picture quality over PAL, SECAM or NTSC standards
Digital sound encoding permits up to eight high-quality sound channels
High-rate digital burst in a flexible sound/data multiplex
Scanning system is compatible with existing television receivers for PAL and SECAM
Future requirements for wide aspect ratio picture are catered for
The ability to accommodate future enhancements in picture quality
Designed to be comptible with the digital studio standard
Provision for conditional access and scrambling. Also suitable for cable

Within the overall data capacity of the system the numbers and types of sound channels can be varied at any time to suit the transmission needs. This is one of the important features of the MAC systems. These features mean that the MAC systems are particularly suitable for European broadcasters. For example, on say, the Eurovision Song Contest, it would be possible to transmit eight different languages of commentary.

Table 10.3 *Summary of MAC/packet systems for television broadcasting by satellite and cable*

C-MAC version	Originally developed to meet the WARC 77 recommendations. Free from artifact deficiencies of PAL, SECAM amd NTSC. Suitable for HDTV, carries eight sound channels. Requires bandwidth of 27 MHz
B-MAC version	Carries six sound channels, teletext, messages, and control data for subscription. The B-MAC system was developed by Scientific Atlanta for its HACBSS
D-MAC version	The version adopted for European broadcasting by the Eureka 95 Committee. Carries eight sound channels, eight commentary channels, teletext and data
D2-MAC version	Half the sound capacity of D-MAC

The D-MAC system can be carried on cable networks where the channel bandwidth is at least 10.5 MHz. For cable networks with a narrower bandwidth (8.5 MHz) the D2-MAC version is more suitable.

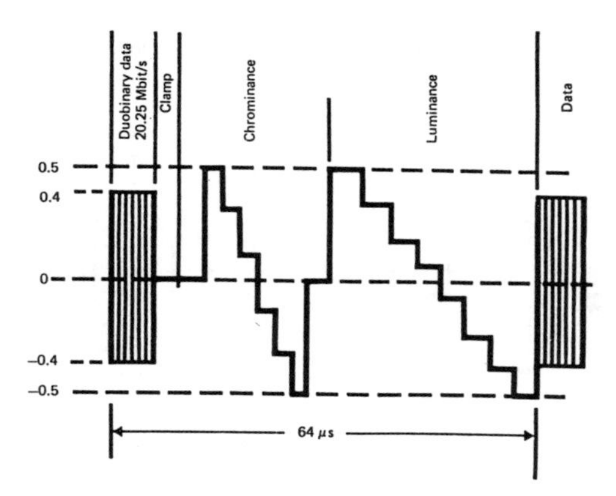

Figure 10.8 Waveform of a D-MAC signal (Courtesy of IBA)

Conditional access and encryption

The basis of the majority of satellite television programmes is that they are on a revenue-earning basis, that is to say they are on a pay-view basis. In order to prevent reception by non-authorized viewers the television programmes are rendered unintelligible by the process known as scrambling. Usually only a proportion of the programmes is treated in this manner; the remainder of the programmes are transmitted in clear for the purpose of attracting new pay-view subscribers. There are other satellite broadcasters who do not scramble any programmes; falling into this category are certain stations operated as a government-subsidized service for cultural reasons. In the final analysis the programme supplier determines whether the programme is transmitted in clear or in a scrambled form.

However, scrambling alone is insufficient if a pay-view format is desired by the channel operator. The television signals may be scrambled but some device is needed to unlock the scrambling process; this is known as encryption. Broadly, the scrambling section is that part which processes the programme whilst the encryption section is that part which processes the key signals which lock or unlock the programme to the viewer.

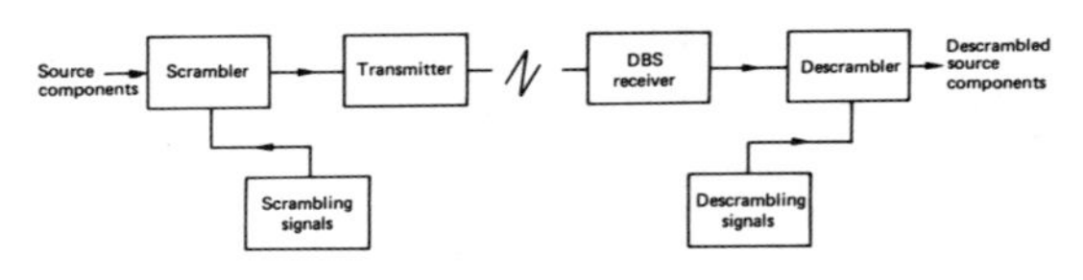

Figure 10.9 The basic scrambling system (Courtesy of IBA)

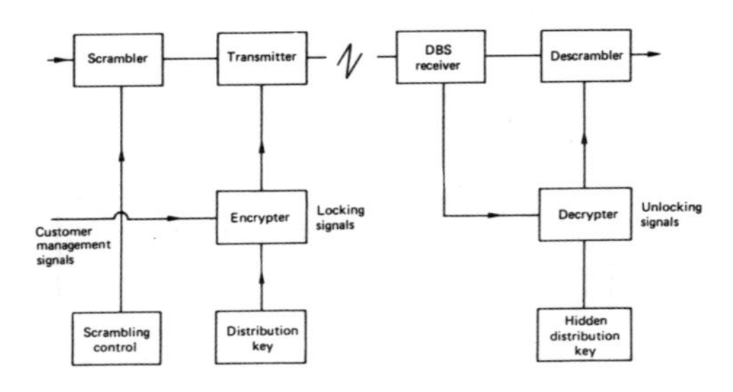

Figure 10.10 The basic scrambling system with addition of an encryption process (Courtesy of IBA)

The D-MAC system has been chosen to be the future standard for most of the European countries and the following description relates to the D-MAC family.

Scrambling

The following is a brief description of the way in which D-MAC operates. For both colour and luminance waveforms, variable cut points are chosen within a determined flight range of 256 equally spaced positions. The exact positions of the cut points vary from line to line and therefore will cause the received signal to be unintelligible to viewers.

Two 8-bit numbers, one for each waveform, are required in order to select the cut positions. These are obtained on a vision line from a pseudo-random generator. Sound and data scrambling is achieved by combining the output from the pseudo-random generator with the useful data of each packet. Both of these scrambling processes are designed to be reversible in the television receiver system by generating an identical series of numbers to those used in the transmitter. This process of recovering the correct signal from its scrambled form is called descrambling.

Encryption

The process of locking and unlocking the scrambling process is called encryption. Key signals to perform this encryption process can be done in a different number of ways but whichever process is used the final objective is to ensure that secrecy is maintained on the exact coding and it must be sufficiently sophisticated to prevent de-encryption by those who seek to pirate reception. The encryption system can deliver the session key to the customer over the air provided the customer is entitled to the service as in the pay-view system.

Figure 10.9 shows the scrambling system in its basic form, and Figure 10.10 shows this scrambling process but with the addition of an encryption process.

Because the sound and vision signals of the D-MAC packet system are divided into discrete samples, the MAC waveforms lend themselves to being easily adapted to scrambling techniques.

A number of proprietary scrambling systems have been developed. From an analysis of satellite transponder channels the ones in most common usage are Eurocrypt M, Eurocrypt S, and VideoCrypt.

11 Propagation of satellite signals

Propagation characteristics

That part of the frequency spectrum allocated to satellites is commonly known as the gigahertz band, and it encompasses all those frequencies from a lower limit of about 1 GHz to an upper limit of 18 GHz – though there are intentions to introduce another band, the Ka-band from 20 to 30 GHz.

All frequencies from VHF upwards are line-of-sight communication but different frequency bands exhibit differing characteristics. At the lower end of this frequency spectrum the noise temperature of the galaxy increases with decreasing frequency, whilst at the upper frequencies the problems are more to do with absorption of microwave energy. This is due to a number of different effects in the atmosphere.

In Europe the main choice of frequencies for satellite broadcasting is the Ku-band; this uses frequencies in the range of 14 GHz for the uplink and 11–12 GHz for downlinking, i.e. the signals picked up by SMATV systems and for individual viewers on a direct-to-home (DTH) basis.

Leaving to one side the effects of absorption due to water vapour, ice crystals and snow, scintillation and other effects for which additional allowances have to be made, the direct path loss on the uplink is of the order of 200 dB and the same amount of path loss experienced on the downlink, thus making an overall path loss of 400 dB. From this fact the basic parameters of a satellite communication system can be defined:

1. The earth station for the uplink should have large gain to counteract path loss; this is achieved by using very large antennas.
2. The receiver in the satellite transponder should have a low noise factor so as not to degrade the quality of the received signal.
3. The transmitter in the transponder should use high-power amplifiers.
4. The on-board antennas in the satellite should have the maximum possible gain, both on the uplink receiver and on the downlink transmitting antenna.

A typical satellite repeater may have an overall gain of 115 dB, with about 54 dB being attributed to the main power amplifier at 12 GHz, i.e. the travelling wave tube amplifier.

Although satellite communication uses frequencies that propagate in line of sight (LOS) as with terrestrial UHF, it is here that the resemblances end. It is the height above earth, 36 000 km, that enables a single satellite in space to communicate with almost half of the surface area. Thus a single satellite in space could broadcast to a landmass the size of Europe. This amount of coverage would not be possible with a terrestrial-based transmitter using UHF, nor would the same coverage be possible with a low earth orbiting satellite, because the angle subtended would be too great for the normal beamwidth of a satellite's antenna.

It is this ability of a satellite in a geosynchronous orbit 36 000 km above the earth that distinguishes satellite communications from terrestrial television transmitters. A terrestrial-based UHF television transmitter system needs hundreds, and in some instances many thousands, of transmitters and transponders to provide a network that will reach 99.9% of the population.

Because of the fundamental limitations of broadcasting television programmes by LOS on VHF and UHF frequencies, this means that signals can rarely travel for more than 70–80 km over a reasonably flat type of terrain, and sometimes for only a few kilometres under hilly conditions. The BBC uses over 1700 transmitters to cover the whole of the British Isles.

An even more graphic example of the limitations associated with terrestrial television is that of Japan. Japan is a country fragmented by nature. For the most part it is composed of hills, valleys, ravines, rivers and mountains. In addition to the natural obstacles to LOS transmission, Japan is also composed of a number of lesser islands. These distant islands, most of which are inhabited and therefore need to receive television programmes, have in the past been prevented from doing this because of the separation distances from the mainland being greater than that possible with LOS due to the curvature of the earth. In addition to all of these natural obstacles in the way of an efficient terrestrial television network, Japanese cities suffer from signal ghosting due to man-made buildings.

It was precisely for these reasons that Japan was the first nation to launch a satellite for DBS, to bring television to countless thousands who had been denied the pleasure.

Satellite communications and DBS systems are not without their propagation difficulties, but they are usually of a predictable nature, and measures can be taken in most cases to ensure that viewers, wherever they may be, with very few exceptions, are able to receive television through DBS systems.

The following are some of the more important factors to take into account:

(a) free space losses
(b) attenuation due to cloud and rain
(c) atmospheric attenuation
(d) Faraday rotation effect
(e) signal losses due to scintillation
(f) adjacent channel interference
(g) interference from terrestrial transmission

Free space losses are a function of the height above earth of the geostationary satellite. The distance is at its minimum when the satellite is directly above the ground receiver, and increases when the satellite is near to the horizon. These typical figures for the three cases, zenith, average and horizon, are plotted in Table 11.1. Though the additional signal loss amounts to only about 2 dB, it is of significance when added to other losses.

Cloud and rain have an effect upon transmission in the gigahertz range of frequencies. The common denominator is

Table 11.1 *Free space loss*

Frequency (GHz)	*d* 35 787 km (zenith)	*d* 39 000 km (typical)	*d* 41 679 km (horizon)
4	195.6	196.3	196.9
6	199.0	199.8	201.9
11	204.4	205.1	205.7
12	205.1	205.8	206.4
14	206.5	207.2	207.9

water droplets, which can become snow or ice. Signal attenuation from this phenomenon, called hydrometeors, increases with frequency. This means that the new Ka-band from 20 to 30 GHz will suffer more than the C-band and Ku-band of frequencies.

Atmospheric attenuation is due to absorption of microwave energy from oxygen, water vapour and rain. Fortunately the frequencies most affected fall into the higher regions from 22 to 183 GHz. The effect of atmospheric attenuation is less serious in C-band and Ku-band frequencies.

The Faraday rotation effect is a phenomenon experienced by signals which have travelled through the ionosphere. These signals interact with the earth's magnetic field to produce an effect known as Faraday's rotation. The effect is proportional to the density of the ionized layer and results in an attenuation of the signal.

Scintillation losses are the result of signals having to pass through several ionized layers which range from 80 to 450 km above the earth. The effect of scintillation in the case of satellite broadcasting is greater when the satellite is near horizon level with consequent low angles of elevation at the receiving installation. Scintillation causes signals to fade. It is subject to time of day and region of the world, and can reach as much as 10 dB signal loss.

Adjacent channel interference is the result of an interfering signal in an adjoining channel. The WARC recommendations for allocating adjacent channels to a satellite serving a different region have gone a long way towards reducing this phenomenon.

Interference from terrestrial transmissions is not normally a problem. However, it is more likely to be found in satellite receiving systems whose dish angle of elevation is near to the horizon. Under these conditions the dish can pick up a lot of ground clutter and interference which would not be received if the dish was pointing towards the sky, i.e. 90°.

A well-engineered receiving system for DBS will take into account all the aforementioned. In addition to the free space loss, which is the sum of the signal attenuation suffered both on the uplink and downlink, the system design should allow for the losses mentioned, and in particular the positioning of the receiving dish in as clear a situation as is possible, and the power delivered in the footprint served by the satellite. This signal density falls off at the outer edges of the footprint. Here it may be mentioned that few satellites, even those presently used for DBS broadcasting, use transponder output powers to the maximum level permitted by WARC. This difference can amount to a signal to noise worsened by as much as 4 dB. Astra satellites, for example, use transponders with 55 W output, but the Olympus satellite uses transponders with four times this power.

A phenomenon which critics of DBS often point to is that of transmission loss due to an eclipse. In practice it is the case that the effect of an eclipse is not all that serious. However, a full description of this is included in this chapter.

A possible cause of trouble in a poorly designed satellite receiving system is that due to polarization discrimination. Circular polarization is used in DBS broadcasting because it effectively doubles the channel allocation capacity by permitting two transmissions on the same frequency. In addition, circular polarization reduces the possibility of adjacent channel interference if the adjacent channel is of opposite signal polarity. This measure is important in satellite broadcasting, firstly because of the much higher transmitted power with satellites than with terrestrial transmitters, and secondly because of the very narrow spacing between a satellite and the next satellite, only two degrees of arc. This restriction on spacing comes about because the equatorial geosynchronous orbit can only accommodate so many satellites.

There are two ratios that are of importance in a satellite receiving system. The first is cross-polarization isolation; this is the ratio of the co-polarized signal to the cross-polarized signal. The second is cross-polarization discrimination, which is the ratio of co-polarized signal to the cross-polarized signal when one polarization only is transmitted.

Interruption of service due to eclipse

It has to be recognized that there are certain natural hazards and effects in satellite communications that cannot always be prevented from occurring. Falling into this category is an eclipse. When this occurs it can disrupt satellite communications.

When certain conditions coincide, that is when the sun, the earth and the satellite are in a straight line, then the satellite passes into the shadow of the earth. The effect of this is to block off the source of solar energy to the solar cells mounted on the body of the satellite. This condition is a nuisance because it causes disruption to the power generated by the solar cells and thereby brings about a loss in transmission service. Satellites of modern design consume as much as 3 kW of electrical power in the transponders and other essential services. To mitigate these effects it has become standard practice to fit a back-up battery power unit, usually of the rechargeable Ni–Cd type, but even then the weight of a 3 kW battery back-up would make it prohibitive and a compromise arrangement does not provide sufficient electrical power to drive essential services to keep the satellite in its correct orbit as well as power the transponder. As a result the power to the transponder is not always sufficient to drive all transponders to saturated power.

Fortunately the phenomenon of the eclipse happens only twice in every year, from 28 February to 13 April, and again from 1 August to 15 October. For most of this time the effect is a loss in satellite communications of several minutes each day, rising to a maximum of 72 minutes at the equinox periods. It is fortuitous that the loss of service always occurs at midnight, true solar time, at the longitude of the satellite, a time when most communications satellites are carrying a low rate of traffic.

In addition to the possibility of a loss of service under eclipse conditions there are two other undesirable, but predictable, phenomena which affect the quality of transmission service to the user. The first of these occurs when the sun aligns with the satellite in the beam of the receiving antenna. When this happens, for periods of about ten minutes on five consecutive days a year, the ground station receives a vast increase of solar noise which causes severe signal degradation. The second phenomenon is that brought about by the eclipse of the satellite by the moon. Fortunately this happens most infrequently; the next serious eclipse by the moon will take place in 1999, and last up to 51 minutes.

Gain and performance calculations for a satellite DBS system

Radiated output power from transponder

The effective output power from a transponder is the product of the output power measured in watts, multiplied by the gain of the transmit antenna, which is always expressed as being relative to an antenna in free space, i.e. isotropic. For this reason the output power is always expressed as the effective isotropic radiated power, EIRP:

$$\text{EIRP} = W_{tx} \times G_t$$

where EIRP is the effective isotropic radiated power, W_{tx} is the output power, at saturated power, from the transponder, and G_t is the gain of the transmitting antenna in dB: relative to isotropic radiation.

Received carrier power at the satellite receiver

The signal received at the satellite receiver is defined by the equation:

$$P(\text{eff}) = \text{EIRP} - (F_{si} + A) - G_r$$

where F_{si} is the free space loss measured in dB, A is the loss due to fading, and G_r is the gain of the receiver dish dBi.

Free space losses are of the order of 206 dB. Should attenuation due to fading be experienced, an allowance of 10 dB is usual. The gain of the satellite receiver dish varies with diameter; a 0.6 m dish will provide a signal gain of about 35 dB.

Free space loss

Free space loss in satellite communications is a function of frequency in use, and the distance the signal has to cover from the satellite above earth and the straight line distance to the receive point. The value of *d*, the distance, is determined by the position of the satellite with respect to the receiving point. Where the satellite is placed in an equatorial orbit, and the receiving point is exactly on the equator, then $d = 35\,787$ km. Where the ground station is on the satellite's horizon, i.e. at a shallow angle, then *d* at its maximum distance = 41 679 km. The typical free space loss for European DBS broadcasting would be that for a typical or average distance of 39 000 km.

Table 11.1 gives the free space loss in dB for these three cases and for a range of frequencies from the C-band and the Ku-band. In practice the losses are always greater than these theoretical figures and further allowances must be made for atmospheric loss L_a, and rain precipitation loss L_p; at 4–6 GHz L_p is quite small, but in the 11–14 GHz range it can be as high as 10 dB during very heavy rain conditions. Atmospheric losses depend upon the elevation of the dish antenna at the ground station, with highest atmospheric losses occurring at low angles of elevation.

Sources of interference in DBS systems

The total electrical noise, which is the unwanted signal, comes from a number of different sources. These can be best illustrated by a drawing of a typical receiving dish (Figure 11.1). The first of the unwanted components is the cross-polarization discrimination, XPD.

The second source of unwanted interfering signals is from satellites in adjacent orbits. The most effective way of reducing and possibly eliminating these interfering signals is by ensuring that the satellite receiving dish is very accurately aligned onto the wanted satellite transmission, as in Figure 11.1.

The two remaining sources of interference are terrestrial transmissions and electrical noise emanating from the earth itself. From the drawing of a typical satellite dish it will be obvious that the higher the angle of elevation the less the risk of any terrestrial interference. Conversely, when the satellite is near to the horizon, the lower the angle of elevation and the greater the risk of interference from terrestrial transmissions, and noise emanating from the earth itself.

Antenna noise temperature is inversely related to angle of elevation (Figure 11.2). The lower the look angle, the higher the noise temperature. It will be seen that this remains fairly flat for elevation angles from 90° to about 30°. Thereafter there is an increase in noise, increasing exponentionally, such that

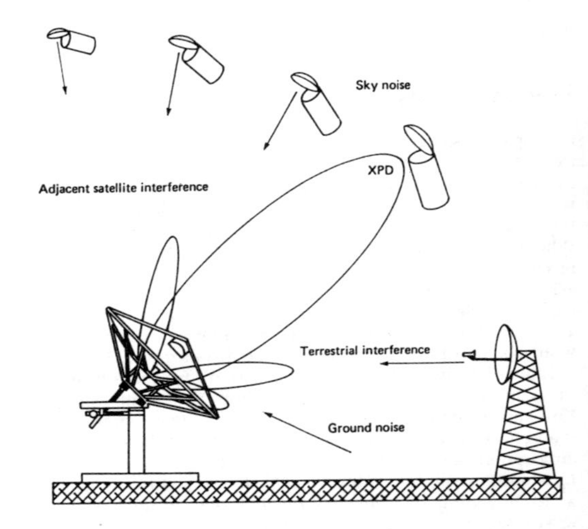

Figure 11.1 Satellite antenna noise comes from several sources. Noise sources include interference from adjacent satellites, the background noise of space, terrestrial interference, the 'warm body' radiation of the earth, and cross-polarity discrimination (XPD) (Courtesy of Broadcast Engineering)

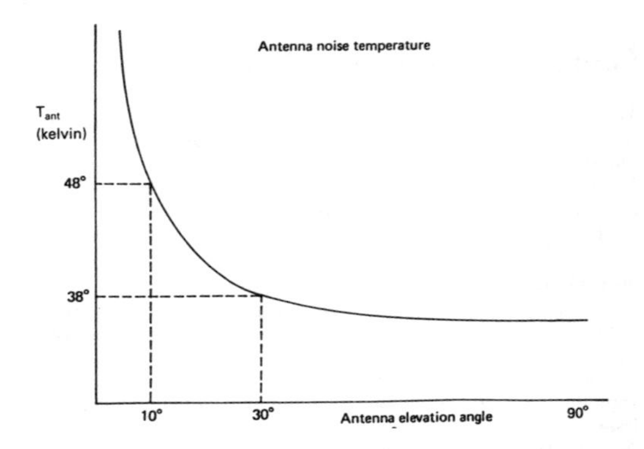

Figure 11.2 Because of the 'warm body' earth, antenna noise temperature is inversely related to elevation. The lower the look angle, the higher the noise temperature (Courtesy of Broadcast Engineering)

10° elevation is the worst signal that should be attempted. At lower angles the noise from both terrestrial and ground sources will make satellite reception very difficult.

12 Domestic receiving systems for DBS

DBS receiving systems

A DBS receiving system must be capable of receiving satellite signals on a DTH basis. Cost must be kept to within certain limits and is therefore possibly the most important parameter because it will determine volume sales. As an example, when the first VCRs were introduced a previous survey had established that reducing the cost from £1000 to £500 would increase sales volume by a considerable factor.

The satellite receiving system must be capable of delivering a picture quality that is equal to, or better than, that received by the television set when it is receiving terrestrial-borne television programmes.

In more technical terms a satellite receiving system should meet or satisfy:

(a) low initial cost
(b) high performance, equal to or better than conventional television
(c) unit interchangeability or replacement at moderate cost
(d) compatibility between DBS receivers and existing television sets
(e) low spurious levels of radiation
(f) protection against interference
(g) be capable of expandability or upgrading
(h) ease of handling as a domestic appliance
(i) the individual units making up the system must be as small as possible

A typical DBS receiving system consists of four basic elements, which are:

(a) the satellite receiving antenna, usually termed 'dish'
(b) an outdoor unit
(c) the interconnecting feeder system to the satellite receiver
(d) the indoor unit, called the satellite receiver

In addition, the viewer needs to have an existing television receiver.

The receiving antenna or dish is installed on or adjacent to the viewer's house or building, usually at a height between 4.5 and 7.5 metres above ground, though there is no upper limit; indeed, the higher the better, if it can be arranged. The determining factor is the need to keep the interconnecting coaxial cable to the shortest possible length so as to reduce feeder losses.

The outdoor unit (Figure 12.2) is physically interposed between the dish and the coaxial feeder; its purpose is to transpose the satellite transmission frequency, usually 12 GHz, down to 1 GHz. The incoming satellite transmission is picked

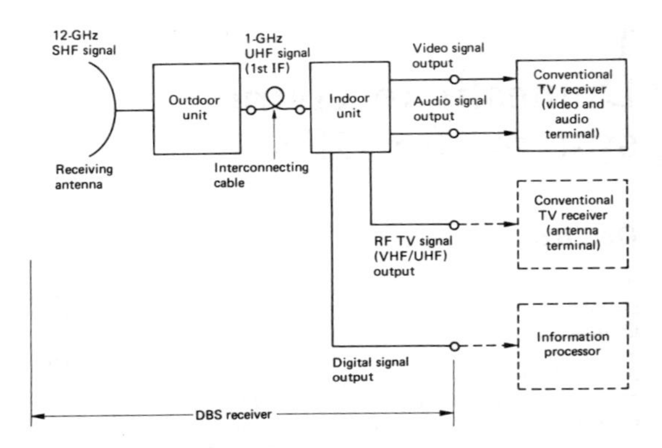

Figure 12.1 Basic configuration of DBS receiver

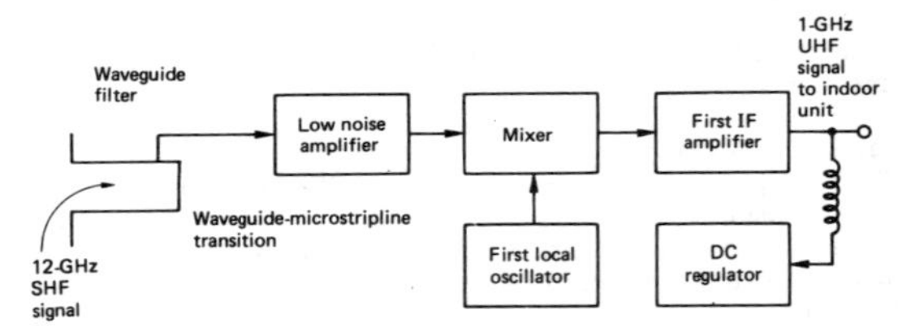

Figure 12.2 Outdoor unit

up by the dish, which has a figure of gain which increases with the size of dish, but for normal DBS transmissions a dish size between 60 and 90 cm is adequate. The signals from the dish are reflected to the primary feed through a polarizer, which converts the polarization to linear.

The indoor unit, the satellite receiver, is mounted in close proximity to the existing television set, usually underneath. This piece of equipment is the only part which has controls. Its main function is to switch or tune from one satellite channel to another. The size and appearance of a typical satellite receiver is similar to that of a VCR. However, this may be regarded as being an interim solution, as by the late 1990s a new generation of television receivers will be in full production. These television sets will incorporate the satellite receiver, but will also be capable of receiving signals from UHF terrestrial-based television transmitters.

An essential feature of this new generation of television receivers must be the capability of receiving television signals in any one of a number of different transmission formats, depending on the part of the world where the receiver is marketed. Additionally they will have much brighter screens, and much larger screens, up to 40 inches diagonal. There will also be a choice of aspect ratios.

The introduction of any new technology, as distinct from normal improvements, is a slow process for two main reasons:

the time taken to develop the new technology to a form acceptable for quantity production, and the need to sell the concept and then the hardware to the consumer. It is not always easy to predict with any degree of certainty how new technology will be accepted by the public and sometimes there are pleasant surprises along the way. If, for instance, satellite television has even half the amount of success enjoyed by VCRs, the manufacturers of satellite systems will be satisfied.

Manufacturers of satellite systems for DBS

Within the short timescale over which satellite broadcasting has been introduced there has emerged an industry that is worldwide, with many manufacturers in the USA, Europe and the Far East. This industry is well supported by area and regional distributors in these parts of the world. The main growth areas of the world are Europe (France, Belgium, Holland, Germany and Great Britain) and the Scandinavian countries (Finland, Norway and Sweden).

Within the past two years there has been a dramatic growth in the retailing of satellite receiving systems, and as a result the consumer is faced with a wide range of products and with an equally wide range of prices. As with any domestic product, the choice can range from the low-cost budget systems sold at an all-inclusive price, up to the opposite end of the market catering for the enthusiastic satellite viewer willing to expend a fairly considerable amount of money on his hobby. Some of these systems are very sophisticated indeed, and can almost be regarded as falling in the professional market, often fitted with very large dishes, fully motorized, and able to sweep the skies.

In between these two extremes there is the middle-priced market catering for the domestic viewer who seeks a high-quality installation but not necessarily one capable of operating as a motorized system. Manufacturers of satellite receivers and systems include the companies shown in Table 12.1.

Table 12.1 *Manufacturers or trade names of satellite receiving systems*

Amstrad	Finlux	Skylink	Triax
Cambridge	Fracarro RI Spa	MIMTEC	Sakura
Chaparral	Grundig	Orbitron	Samsung
Armstrong	PACE	Uniden	Winnersat
Drake	NEC	Fuba	Toshiba
Echostar	Nokia	Triad	Philips

Developments in satellite receivers

As with any new or emerging technology the pace of technical development is fast. The domestic satellite receiver is one example: today a high quality domestic satellite receiver will have a performance that far exceeds the products of the late 1980s. At the 1993 Cable and Satellite Exhibition manufacturers announced new ranges with many new features capable of receiving C, Ku or S band signals, with channel selection capability running to over 100 channels. Other features include automatic antenna tracking, enabling the operator to receive any satellite in the Clarke belt operating in geostationary orbit. Other receivers permit reception of satellites in inclined orbit, a rather more complicated form of tracking in two-axis.

Orbitron Corporation, a well established company in the professional market, has entered the competitive domestic market with its new range of receivers. The 6000E is a single band receiver capable of receiving by switch selection either C, Ku or S band, with bandwidth for video from 9, 18, 27 and 32 MHz. It includes a built-in antenna positioner that can track any satellite in the Clarke belt. The 7000E is an enhanced version of the 6000E. With dual IF inputs and with the addition of an external switch the receiver can be set to receive C, Ku or S band transmissions. The 7000E is pre-programmed to receive all European and Middle East channels from Intelsat 27.5° W, to Asiasat at 105.5° E.

The European division of another major international company, the Echosphere Corporation has announced what is claimed to be the most advanced satellite receiver on the market today. This is its Echostar SR-8700. The SR-8700 is programmed to receive over 1500 channels and can be pre-programmed for 200 video channels. Amongst other features it includes four language on-screen graphics and Wegener quality Stereo. Antenna tracking is a built-in feature. The Echostar SR-8700 has a built-in Videocrypt or D/D2-MAC decoder that can receive high quality video in Videocrypt, Eurocrypt M or Eurocrypt S.

Amongst its very extensive range of satellite receivers Echosphere has a dual-axis, auto-tracking system specially designed to track those satellites operating in inclined orbits. The NOMAD 11 Auto tracker features a pre-programmed facility of 55 orbital positions and the opportunity to custom name five additional satellites. Echosphere has also introduced the NOMAD 1 which is a software controlled dual axis positioner that permits precise antenna positioning in two-axis.

The Echosphere Corporation is not only a manufacturer of satellite technology: Echosphere will be one of America's DBS broadcasters when it launches the first of its seven high power DBS satellites in late 1995. Seven high powered GE Astro satellites Series 7000 are already under construction in a multi-million dollar contract awarded to GE Astro by Echostar Satellite Corporation.

Reception of D-MAC transmissions by DBS

All receivers for DBS require a dish antenna of specified dimensions to suit the prevailing reception conditions, together with an associated head-amplifier/down-converter known as an LNB. The dish must be mounted outdoors, preferably at a high level, and must have an unobstructed view towards the satellite(s). The 12 GHz satellite signals are down-converted to an intermediate frequency and fed by coaxial cable to the D-MAC television receiver or the set-top converter box.

Low noise converter blocks

The LNB is a crucial element in a satellite receiving system. In principle it is a frequency converter and its main function is to take the incoming satellite frequency output from the receiving dish and convert this to an intermediate frequency. Most LNBs accept signals at either the C-band or the Ku-band and convert this to a frequency within the range 950–1750 MHz. Generally, the gain from an LNB is of the order of 45–50 dB, but with a requirement that variation within any of the 27-MHz segments must be held to ±1 dB.

Improvements in the performance of LNBs have been a natural consequence of developments in FET (field effect transistor) technology. Table 12.2 is a specification for a modern LNB. These performance figures relate to the Lenson Heath 500 series manufactured by STC.

The layperson sometimes assumes that because the satellite is up in the sky, anyone can receive a satellite transmission, given a suitable satellite receiving system. Unfortunately, this assumption is sometimes incorrect. To receive a satellite transmission the receiving dish must have a clear and uninterrupted optical view of the satellite. This cannot always be guaranteed. Reception problems are more likely to happen in heavily built-up areas, where the obstruction often takes the form of a nearby office block or flats. The actual geographical location of the receiving point has an influence. As an example, the angle of elevation required for a dish in the south of England might be 28°, but to receive the same satellite transmission in the very north of Scotland the dish would have to be at a shallower angle which might be 8° lower.

Table 12.2

Input frequency	10.95–11.70 GHz
Output frequency	950–1700 MHz
LO frequency	10.00 GHz ± 3 MHz
Total noise figure	1.3 average, 1.5 maximum
Conversion gain	46–58 dB
Ripple in 27 MHz band	±0.5 dB
IM3	±10 dB
Image rejection	−80 dB
LO leakage (input)	−80 dB
LO leakage (output)	−25 dB
Cross-polar discrimination	18 dB minimum
Output VSWR	2:1 maximum
Output impedance	75 Ω
Power requirement	11.8–13.4 V vertical, 16.4–18 V horizontal
Current	200 mA

Note: LO = local oscillator, VSWR = voltage standing wave ratio

The height of the dish is an important factor in alleviating problems with siting, so as to clear obstructions on the skyline. However, this positioning needs to take account of the fact that a long feeder length from the dish to the satellite receiver will introduce greater losses.

Receiving dishes for reception of DBS

The dish for receiving DBS broadcasts is performing the same function as the receiving dishes for an earth station or for an SMATV system, with the important difference that its gain is very much smaller; this is one of the reasons why satellites used for DBS need to have much more powerful transponders than do communication satellites. Clearly it would be neither practical nor desirable for domestic satellite television installations to have large, even unsightly, dishes in domestic-type environments.

The WARC recommendations first envisaged domestic-type dishes as not exceeding a diameter of 90 cm, and transponder output powers from the DBS possibly reaching 240 W.

Table 12.3 *WARC specification for a DBS receiving installation*

Antenna	
Type and gain	Type: 60 / 75 / 90 / 100 / 120 (dB): 34.5 / 36.5 / 38.0 / 39.0 / 40.0
Output VSWR	< 1.3
Output port	WRJ-120 waveguide/BRJ-120
Outdoor unit	
Input port	WRJ-120 waveguide/BRJ-120 flange with waterproof
Noise figure	<4 dB
Input signal level	−80 dBm ± 10 dB/channel
Input VSWR	>2.5
Overall gain	48 ± 1 dB
1st local oscillator frequency	10.678 GHz ± 1.5 MHz (−30°–50°C)
leakage power	< −30 dBm
1st IF	1.036–1.332 GHz
Indoor unit	
Input signal level	43 dBm $^{+15\ \text{dB}}_{-18\ \text{dB}}$
Input VSWR	<2.5
2nd IF	either 134.26 or 402.78 MHz
Leakage power of 2nd local oscillator	< −55 dBm
APC gain	within ±500 kHz against ±2 MHz deviation
Received signal quality	
Frequency response	video: +1 dB for 50 Hz 4.2 MHz, within −3 dB at 4.6 MHz sound: within +1 − −3 dB at 15 kHz for A mode within +1 − −3 dB at 20 kHz for B mode
Linearity	video: Differential gain <5% Differential phase <5°
Bit error rate	$<3 \times 10^{-4}$ at C/N = 9 dB, before error correction

(IR, Study Group. DOC, 10-11S/J-12, Japan)

However, since then developments in Ku-band technology have exceeded the early theoretical expectations, with the result that it is now possible to operate a satisfactory DBS system with lower output power from the transponder, and with smaller diameter dishes at the viewers' homes.

There have been some corresponding developments in dish technology and there are now several major companies producing a wide range of dishes. There are three design types of dishes, each having its merits. These are:

(a) the centre feed dish
(b) the offset feed dish
(c) the flat plate antenna

The centre feed dish

The centre feed dish is structurally superior to other types, its main disadvantage being that there is a loss in electrical efficiency due to signal blockage. The structural advantages of this method mean that it is suited to very large earth station dishes. There is a way of reducing signal blockage by the use of the so-called button hook mount. This arrangement is suited to medium-sized dishes, i.e. 3–4 m.

The offset feed dish

This is easily the most popular of all types for applications in DBS. It is simple in construction and also in dish fabrication, since it is a single-piece construction made either by pressing, spinning or moulding. When the one-piece reflector dish is made, it is usually dip-coated. An important part of any dish design is that it should be of smooth construction so as to have low retention of snow.

Three of the most important parameters in performance are the efficiency, the cross-polar discrimination (XPD) and the gain. It is the figure of gain that varies with dish size, and hence it is important to select a size of dish suitable for a particular satellite. Generally the gain may be expected to be roughly as follows for different dish size: 45 cm, gain 33.39 dBi; 55 cm, gain 35.1 dBi; 76 cm, gain 38 dBi; 120 cm, gain 42.4 dBi.

The flat plate antenna

The flat plate antenna operates on an entirely different principle to the offset feed antenna. In this arrangement the signals received are reflected by the parabolic reflector focus at the top of a primary feed. These signals collected by the primary feed are then introduced into the outdoor unit. In the case of the flat plate the principles are quite different.

The flat plate antenna has several advantages and may well prove to be popular in the long term. It is simple to mount, has low wind resistance, has no feed blockage, and is able to handle all polarizations. A flat plate of 350 mm square gives a gain of about 31 dBi, i.e. the same as a 400-mm circular parabolic dish. Because of the complexities in manufacture the flat plate is likely to be more expensive than the parabolic dish type.

Table 12.4

Diameter (metres)	2.5	3	3.5
F/D ratio	0.36	0.36	0.36
Focal distance (cm)	90	90	90
Gain at 12 GHz	46.6	48.2	49.5
Gain at 11.7 GHz	45.9	47.5	48.8

(Courtesy of Orbitron Corporation)

Dishes for SMATV

Satellite master antenna systems are used to deliver television programmes via satellite to many viewers. Because of this the dish needs to be very much bigger so as to have the required gain. One of the foremost companies manufacturing such dishes is Orbitron (Table 12.4). This company has a standard product range covering dish sizes from 2.5 to 3 metres. Gains vary from 46.6 to 48.3 dBi at 12 GHz.

Every Orbitron antenna features a pre-assembled polar mount with spinclination declination adjustment. This feature permits one person to assemble to the desired bearing in azimuth and elevation.

For even bigger installations such as sub-earth stations, Orbitron manufactures another range. This is the motorized version, available in two standard dish sizes. These are suitable for the C-band and the Ku-band, with the performance shown in Table 12.5 for the main parameters.

Table 12.5

	T-20 (6.1m)	*T-24* (7.2 m)
F/D ratio	0.45	0.375
Beamwidth at 4 GHz	0.86°	0.73°
Beamwidth at 12 GHz	28°	24°
Gain at 4 GHz (dBi)	46.1	47.8
Gain at 12 GHz (dBi)	55.1	56.5
Efficiency 4 GHz (%)	68	68
Efficiency 12 GHz (%)	55	55

(Courtesy of Orbitron Corporation)

13 Satellite television as a mass medium

BBC World Service Television

BBC World Service Television is a new channel for a new age in international broadcasting. The service first began in March 1991 under the flagship of BBC World Service Television Limited.

BBC WSTV inaugurated its satellite service in March 1991, and moved to 18 hours per day Monday to Friday, and 12 hours during weekends. All programmes were then transmitted in the English language only, and were available throughout Europe on a subscription-only basis, direct to homes, hotels and cable companies. Its biggest subscribers were the hotel chains, providing a service to its business customers in all bedrooms.

For this first service the BBC signed up with Comsat for satellite Intelsat VI at 332.5°E spot beam. All that is needed to receive the BBC service from Intelsat VI is the decoder, which can be purchased from BBC World Service Television, Woodlane, London. Having once purchased the decoder, thereafter a small annual contribution is payable to the BBC. The east spot beam of Intelsat VI (Figure 13.1) is centred over Lyon, France, and for the footprint area over Western Europe an 80 cm dish will suffice. The footprint extends as far as Moscow but for this a 2.5 m dish would be needed. The signal encryption uses the SAVE system but the BBC says it intends to change to a more sophisticated system by the end of 1993.

The programme content of BBC WSTV is essentially news-based, designed to appeal to an international audience. However, to retain the interest of viewers it will be interspersed with documentaries, features, weather reports for the continent of Europe, as well as informed political comment. In essence, BBC WSTV will become the visual arm of BBC World Service.

One other important benefit that Britain enjoys from the BBC World Service is the importance that the BBC attaches to teaching English by radio, and now television. The ability to speak and understand the English language is appreciated by over 100 countries whose peoples realize the value of English as a route to securing better job prospects. BBC WSTV will continue this tradition of teaching by radio.

In October 1991 BBC WSTV extended their satellite service to 38 Asian countries. The fact that the BBC chose to target Asia for this new service underlines the importance that this part of the world has attained in recent years. BBC WSTV is broadcast on one of the five channels occupied by STAR-TV on the AsiaSat 1 satellite positioned in a geosynchronous slot above Borneo. AsiaSat 1 has two beams that provide footprints over most of central and southern Asia; in fact, they cover an area of the globe that extends from the Red Sea in the Middle East to the Yellow Sea in China.

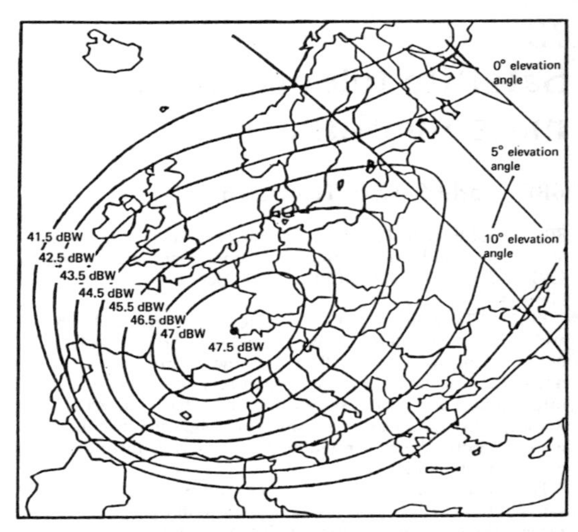

Figure 13.1 Intelsat VI at 332.5°E east spot beam. Nominal power contours assuming one video channel per transponder (present Intelsat VA pointing). Transponder 1E, 10.995 GHz, audio 6.65 MHz, vertical polarization. Mode of encryption: SAVE. Video and audio scrambled (Courtesy of BBC WSTV)

The southern footprint (Figure 13.2) transmits in PAL format since this is the most commonly used format for that region.

The northern footprint (Figure 13.3), covering northern and eastern Asia, transmits its programmes in the NTSC format, and for the same reason, i.e. it is the most commonly used format for that part of Asia.

The EIRP from both footprints hardly comes within the category of DBS requirements; they are a maximum of 36 dBW in the primary region of footprint 1 (northern) and a minimum of 28 dBW at the outer fringes, and figures of 36 to 23 dBW respectively for footprint 2 (southern). This means that dish sizes will have to be large. In fact, BBC WSTV recommends a dish size of 2.4 m in Hong Kong when using a domestic satellite receiving system, rising to a dish size of 3.7 m for a SMATV system where a greater signal to noise ratio is required. In the cities of Peking, Shanghai, Bangkok, Delhi and Kabul the recommended dish sizes are 1.8 and 3 m respectively for domestic television and SMATV. In Saudi Arabia, where the footprint is worst case, the recommended dish sizes are 6 m and 7 m for domestic television and SMATV respectively. These sizes are quite large, calling for professionally engineered installations (Table 13.1).

Nevertheless, it should be borne in mind that satellite DBS has not caught on in the same way as it has in Europe; wealthy citizens in Asia can well afford to invest in professional satellite systems, which require very large dishes. This investment by the BBC in a Far East satellite service is part of a long-term plan based on a forecast that the service

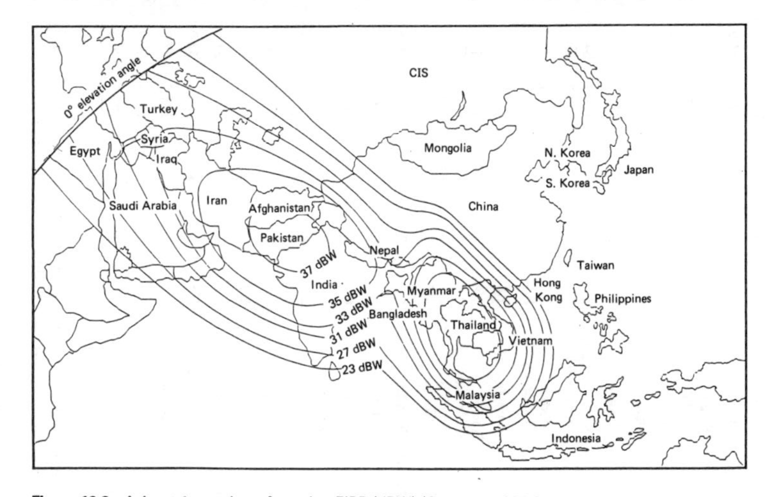

Figure 13.2 Asiasat 1: southern footprint, EIRP (dBW) (Courtesy of BBC WSTV)

Table 13.1 *Recommended antenna sizes for particular locations within Asiasat 1's footprint are as follows:*

		Suggested Size (m)			*Bearing*	
Country	*Location*	*TVRO*	*SMATV*	*CATV*	*Az*	*El*
Japan	Tokyo	3.7	6	8	229.3	35.1
	Osaka	3.7	5	8	225.4	39.0
	Sapporo	3	3.7	5	227.3	29.4
Hong Kong		2.4	3.7	5	201.9	62.0
Taiwan	Taipei	2.4	3.7	5	215.0	55.4
Philippines	Manila	5	7	9	227.0	64.9
China	Beijing	1.8	3	3.7	196.1	42.0
	Shanghai	1.8	3	3.7	208.6	49.9
Korea	Seoul	2.4	3.7	5	213.2	41.5
Mongolia	Ulan Bator	2.4	3.7	5	182.0	35.4
Thailand	Bangkok	1.8	3	3.7	160.1	72.6
Malaysia	Kuala Lumpur	2	3	3.7	127.2	84.0
	Johor Baharu	3	3.7	5	130.8	87.3
Indonesia	Medan	3.7	5	8	115.1	80.8
	Palembang	5.5	7.5	10	165.9	86.4
Pakistan	Karachi	1.8	3	3.7	119.0	40.0
India	Delhi	1.8	3	3.7	131.7	44.9
	Calcutta	3	4.5	6	141.0	57.3
	Bombay	2	3	4.5	120.9	51.4
	Madras	3.7	6	8	115.5	57.0
Saudi Arabia	Riyadh	6	7	9	103.9	19.8
Israel	Tel Aviv	3.7	6	3	100.3	7.4
Egypt	Cairo	6	8	9	97.9	4.7
Turkey	Ankara	6	8	9	101.0	4.4
Syria	Damascus	3.7	5	8	103.6	11.1
Iraq	Baghdad	3	3.7	5	106.7	16.4
Iran	Tehran	2.4	3.7	5	112.6	20.7
Afghanistan	Kabul	1.8	3	3.7	127.8	34.0
Nepal	Kathmandu	2.4	3.7	5	146.1	52.0
Vietnam	Ho Chi Minh	3	3.7	5	185.0	77.7
Laos	Vientiane	3	3.7	5	170.8	68.7
Cambodia	Phnom Penh	3	3.7	5	175.5	75.4

will be delivered to viewers in the major cities of Asia through existing terrestrial networks, satellite cable networks and MMDS.

BBC WSTV programmes appearing in Asia are a result of a joint venture between the BBC and Satellite Television Asian Region, identified as STAR-TV. STAR-TV is itself a 50–50 joint venture between the Hutchinson Group and a Hong Kong business family. The origin of STAR-TV itself is an indication of the way in which satellite television is expected to forge ahead in Hong Kong and the rest of Asia. The service is transmitted free with no encryption, and available to homes with their own satellite dishes, but the main outlets are the network operators delivering by one or other of the delivery systems mentioned earlier. In the main the target audiences are the well-educated, internationally orientated, and well

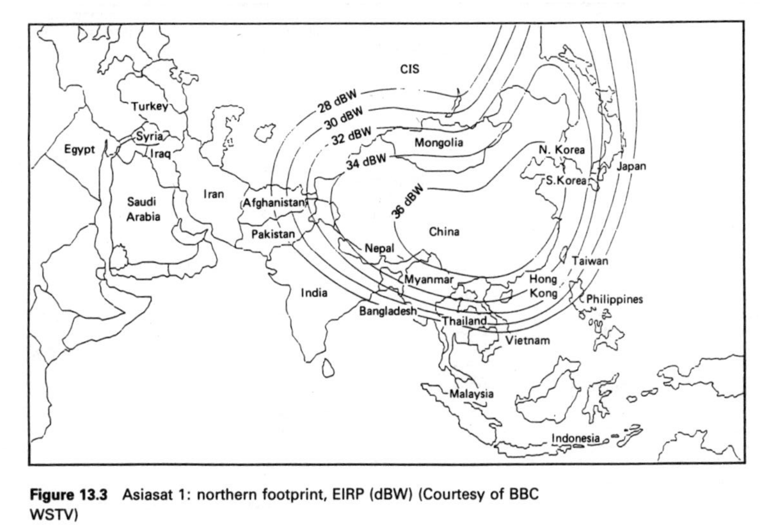

Figure 13.3 Asiasat 1: northern footprint, EIRP (dBW) (Courtesy of BBC WSTV)

Table 13.2 *Frequencies for STAR-TV's programming:*

Programme	*Frequency*	*Beam*	*Standard*	*Polarization*
Sports Channel	3800 MHz	North	NTSC-M	Horizontal
	3860 MHz	South	PAL-B	Vertical
Music Channel	3840 MHz	North	NTSC-M	Horizontal
	3900 MHz	South	PAL-B	Vertical
News/Information Channel	3880 MHz	North	NTSC-M	Horizontal
	3940 MHz	South	PAL-B	Vertical
Mandarin Channel	3920 MHz	North	NTSC-M	Horizontal
	3980 MHz	South	PAL-B	Vertical
Entertainment Channel	3960 MHz	North	NTSC-M	Horizontal
	4020 MHz	South	PAL-B	Vertical
Video Signal Parameters (Southern Beam)				
Television standard:	PAL-B 625 lines, 25 frames per CCIR Report 624-4			
Pre-emphasis:	CCIR Recommendedation 405-1, 625 lines			
Maximum video bandwidth:	5.0 MHz			
Peak-to-peak deviation:	20.0 MHz			
Video Signal Parameters (Northern beam)				
Television standard:	NTSC-M 525 lines, 30 frames per CCIR Report 624-4			
Pre-emphasis:	CCIR Recommendation 405-1, 525 lines			
Maximum video bandwidth:	4.2 MHz			
Peak-to-peak deviation:	21.6 MHz			

informed. These are the same groups of people with spending power, and hence STAR-TV is a commercial broadcaster.

An examination of BBC WSTV indicates the way in which news has assumed the status of a highly marketable commodity. Important as it is to the viewer, it is of greater importance to those who make it and disseminate it. News does more than inform; it influences the way people think and behave. It can even influence their political beliefs. The potential for the BBC World Service to project the British influence around the world is immense. Asia TV has a capability of reaching out to an audience of 2.3 billion people in Asia; that is half the population of the entire world. Given such an opportunity, it is likely that BBC WSTV will in the future outstrip the performance of the traditional BBC World Service programme.

STAR-TV

The recommended specifications for equipment used to receive STAR-TV transmissions on the AsiaSat 1 satellite given in Table 13.2, also serve as the criteria for STAR-TV's equipment and receiving system endorsement programme.

RF Transmission Parameters

STAR-TV's five channels occupy ten transponders on AsiaSat 1, a C-band telecommunications satellite in geosynchronous orbit at 105.5 degrees east longitude. It carries a total of 24 transponders, each being 36 MHz wide. AsiaSat 1 has two beams, one covering northern and eastern Asia, the other covering from the south-east towards the west up to the satellite's horizon in the Middle East. Linear polarization is used – horizontal on the northern beam and vertical on the southern beam.

Worldnet

Voice of America is the international broadcasting arm of the United States Information Agency (USIA), which is its parent body. VOA is funded entirely by the US government through USIA budgets.

In May 1984 the USIA recognized the coming importance of satellite broadcasting by its setting up of Worldnet.

Worldnet reaches out to viewers from its Washington studios, via:

(a) three international satellites
(b) two regional satellites
(c) two domestic satellites

The primary networks start out from Washington, and are uplifted to the Atlantic Ocean relay satellite (AOR), the Pacific

Ocean satellite and the Indian Ocean satellite. In Europe the three regional satellites are European domestic satellites ECS-F1 and ECS-F2, and Telecom 1.

The function of Worldnet is as a mass-media television network operating on a global basis. It does this by transmitting recorded and live programmes from its studios in Washington DC to foreign reporters around the world. Worldnet, in addition to generating its own programmes of news, special features and comment, also acts as a news-gathering agency.

Arabic language broadcasting by satellite

Great Britain and the USA are not the only countries which have acknowledged the merits of satellite television as a suitable vehicle for information broadcasting. In October 1991 the Middle East Broadcasting Centre launched an international language satellite television station. From its London headquarters MBC will broadcast across the whole of Europe via the Eutelsat II F1 satellite. For the coverage of the Arab world in North Africa, the Middle East and the Gulf region, MBC will use satellites belonging to Arabsat.

This Arabic service is the first of its kind, providing information and special interest programmes to a potential audience of over 5 million across Europe alone, and another 100 million viewers in 22 Arab countries. Though MBC is news and information led, the programmes also include some tailor-made for the Arabic-speaking viewer. Like CNN, MBC hopes that its programmes will be marketed to cable television operators, the hotel chains across Europe, apartment blocks where the residents are chiefly Arabs and also to those Arabs who are resident in Europe with their own satellite television receiving installations.

14 Astra, Eutelsat and CNN

Astra

Societé Européenne des Satellite (SES) is the privately owned Luxembourg-based satellite broadcasting consortium in which the Luxembourg government has a 20% holding. SES owns and operates the Astra satellites, and has three Astra satellites in operation. The first was launched on 11 December 1988 from an Ariane 4 launch vehicle and at that time it was one of the most powerful, with 16 transponders plus 6 spares, each having an output power of 45 W. The life expectancy of the satellite is 12.1 years.

The second satellite in the series was launched on 22 March 1991. Once again the satellite manufacturer was GE Astro Electronics, but with two main differences; the expected life is calculated at 14.5 years, and the output power from the 16 transponders was increased to 60 W.

Before 1995 SES intends to launch two more Astra satellites. Like Astra 1A and 1B (Figures 14.1 and 14.2) the next two satellites will be positioned at 19.2°E. Even by the most modest of claims SES must be the most successful satellite broadcasting company in Europe. It is a matter of regret by many that SES is not broadcasting D-MAC programmes, but SES has stated its intention to adopt the D-MAC European standard on the 1D satellite when it is launched in 1994–1995. By so doing Astra will have bypassed the interim HDTV standard of D2-MAC.

The contract to build Astra 1C and 1D has been awarded to Hughes Electronics Corporation in El Segundo, California, and the satellites will be launched from Ariane 4 ELVs. Satellites 1C and 1D will have the same cubic body structure as 1A and 1B and will be fitted with three-panel solar arrays, twin dual polarized antenna dishes and 18 transponders of 63 W output power.

When Astra 1D is launched in 1994–1995 SES will, in addition to having 48 transponders beaming PAL down to the countries of Europe, also be one of Europe's biggest broadcasters of D-MAC HDTV programmes from its 16 transponders on Astra 1D.

In 1992 SES awarded a further contract to Hughes for a new satellite, 1E, scheduled for launch in 1995. When 1E is brought into service SES may sell off the ageing 1A which has been in service since December 1988. Astra 1D and 1E will be the HS 601 versions from Hughes. When these are launched SES's investment in Astra will have reached nearly 1.5 billion US$.

Technical data on Astra 1A and 1B

Astra 1B, the most recent of the Astra satellites, is an advance on the 1A model. It has a wingspan of 24.4 m and a payload on launch of 2550 kg, with an orbit payload of 1246 kg. These figures compare with the 1A payload figures of 1812 kg and 1042 kg respectively. The increased power output from 1B's transponders gives it a larger footprint, which takes in the Canary Islands.

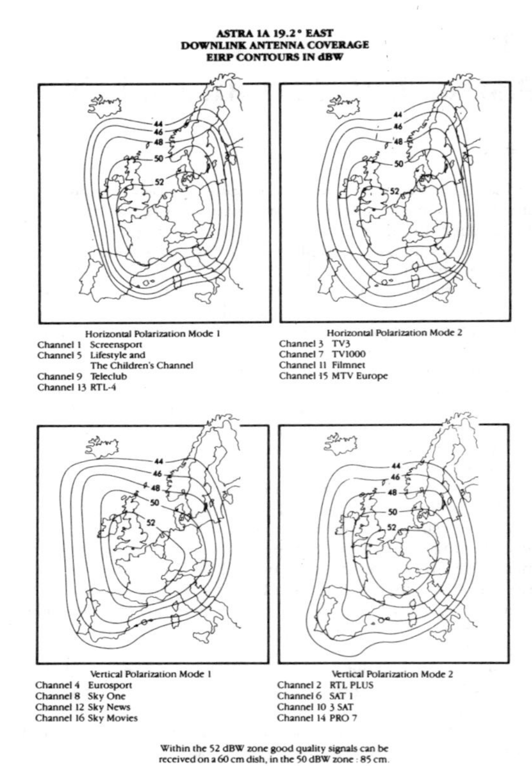

Figure 14.1 Astra 1A 19.2°E downlink antenna coverage; EIRP contours in dBW (Courtesy of Astra)

Growth in satellite cable television and DBS systems by Astra

Astra is Europe's largest satellite broadcaster on a DBS direct-to-home (DTH) basis. It also has a very substantial market in delivering its programmes to viewers through SMATV and cable networks. Statistics for satellite broadcasting should take into account all these modes of delivery.

Astra's approach to the vast potential in the European market was to create a cluster of channels suitable for a particular country. The problem is not as difficult as it may sound, because in many cases all that needs to be changed is the language; in the case of sport programmes, for example, this is fairly easy. Though Astra had met with much success in many European countries with SMATV and cable, it was the UK market that was its big success. In other countries, such as Belgium, SMATV and cable had been established from the

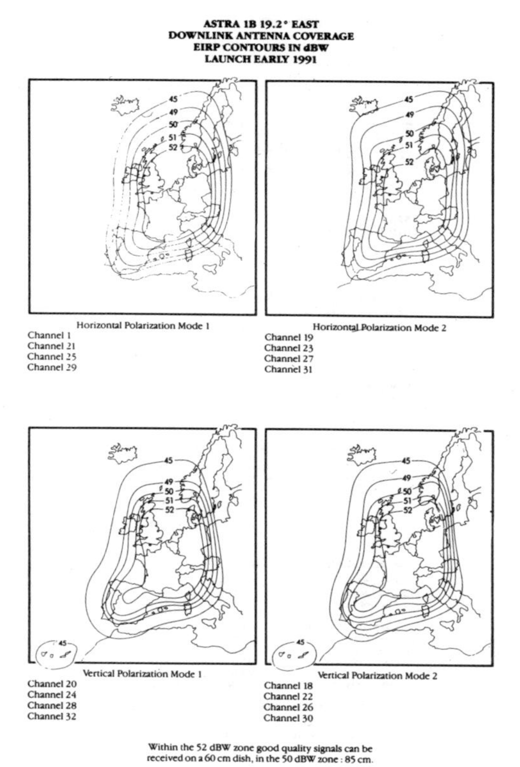

Figure 14.2 Astra 1B 19.2°E downlink antenna coverage; EIRP contours in dBW (Courtesy of Astra)

early 1980s, and today it has the highest penetration in Europe, particularly in towns and cities whose infrastructures lend themselves to relatively low cost cabling.

For DBS, Great Britain is the largest market with Germany a close second. Germany may well emerge as the largest market.

Table 14.2 shows the forecast for the decade 1991–2000 of Astra penetration by both cable–satellite from SMATV and by DBS in Great Britain.

Eutelsat

Eutelsat is the acronym for the European Telecommunications Satellite Corporation, whose headquarters is at 33 Avenue du

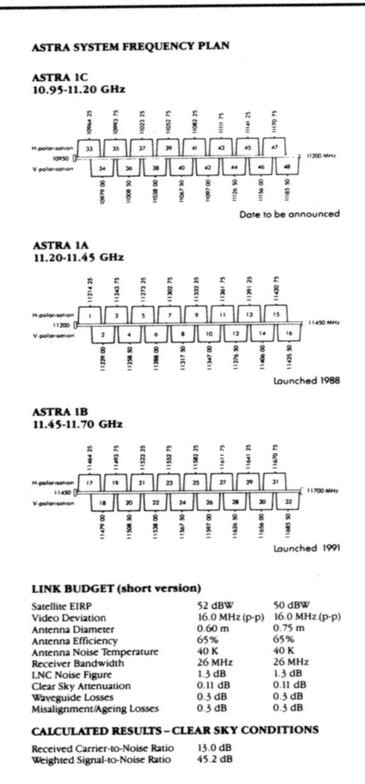

LINK BUDGET (short version)

Satellite EIRP	52 dBW	50 dBW
Video Deviation	16.0 MHz (p-p)	16.0 MHz (p-p)
Antenna Diameter	0.60 m	0.75 m
Antenna Efficiency	65%	65%
Antenna Noise Temperature	40 K	40 K
Receiver Bandwidth	26 MHz	26 MHz
LNC Noise Figure	1.3 dB	1.3 dB
Clear Sky Attenuation	0.11 dB	0.11 dB
Waveguide Losses	0.3 dB	0.3 dB
Misalignment/Ageing Losses	0.3 dB	0.3 dB

CALCULATED RESULTS – CLEAR SKY CONDITIONS

Received Carrier-to-Noise Ratio	13.0 dB
Weighted Signal-to-Noise Ratio	45.2 dB

Figure 14.3 Astra system frequency plan (Courtesy of Astra)

Maine 75755 Paris. Eutelsat is the result of an international agreement between 28 member nations. The organization was formed in 1977 to operate satellites for fixed and mobile communications between its member nations and for these countries' requirements outside Europe.

The satellites have an important degree of flexibility in operating modes. Each can operate in the 12.5–12.75 GHz band simultaneously. In terms of transponder power, capacity and flexibility, Eutelsat can claim to have the largest fleet of satellites for television and radio in Europe.

Even with only four Eutelsat I satellites in orbit, it was the largest operator in Europe. Each Eutelsat I can operate 10 transponders at any one time and provide coverage unmatched by any other satellites in use. Each of its Eutelsat I satellites is equipped with three spot beams which can provide

Table 14.1 *Astra system technical information*

Astra 1A and 1B	
Orbital information	
Orbital location	19.2°E
Stabilization system	Three axis type
Stationkeeping accuracy	±0.1° (E/W and N/S)
Transponder information	
Channel capacity	16 TWTAs plus 6 spares
Nominal EIRP	52 dBW
Eclipse protection	All 16 channels
Transponder bandwidth	26 MHz
Audio subcarriers	Presently 6.50 (or 6.60), 7.02, 7.20, 7.38, 7.56, 7.74 and 7.92 MHz
Astra 1A	
Launch information	
Launch date	11 December 1988, 00h33 GMT
Launch vehicle	Ariane 4, Flight 27
Launch site	ELA2, Kourou, French Guiana
Launch mass	1820 kg
Satellite information	
Satellite manufacturer	GE Astro Electronics
Type	4000
Expected lifetime	12.1 years as from January 1989
TWTA output power	45 W
Astra 1B	
Launch information	
Launch date	2 March 1991, 23h36 GMT
Launch vehicle	Ariane 4, Flight 42
Launch site	Kourou, French Guiana
Launch mass	2620 kg
Satellite information	
Satellite manufacturer	GE Astro Electronics
Type	5000
Expected lifetime	14.5 years
TWTA output power	60 W

targeted coverage of the continent of Europe. The west spot beam covers central Europe, taking in France, Great Britain, Ireland, Benelux, western Germany, part of the Iberian peninsula, Italy and Scandinavia.

The east spot beam takes in Greece, Turkey, parts of Italy and all of eastern Europe, whilst the Atlantic spot beam takes in the Iberian peninsula, the Atlantic islands, and parts of North Africa. For all spot beams the quality of reception is assured with small dishes.

From the launch of its first satellite in 1983, Eutelsat opened up a new era in the provision of television and radio channels for a pan-European audience. For the first time television and radio channels were able to reach simultaneously homes throughout the whole of Europe, from Reykjavik to Ankara. Since its inauguration Eutelsat has extended its reach to many viewers through cable systems in Europe The first of the second generation Eutelsat IIs was launched in August 1990.

With close on a hundred transponders in service, Eutelsat can justifiably claim to be the largest satellite operator in Europe. The operating and design characteristics of the Eutelsat I and II series of satellites are summarized in Table 14.3.

Table 14.2 *1990 forecast for the penetration by Astra in the UK (millions of viewers)*

Year	*By cable and SMATV*	*By DBS*	*Total*
1990	0.3	1.1	1.4
1991	0.7	1.9	2.6
1992	1.1	3.0	4.1
1993	1.6	4.1	5.7
1994	2.0	5.1	7.1
1995	2.5	5.7	8.2
1996	3.0	6.5	9.5
1997	3.2	7.0	10.2
1998	3.5	7.4	10.9
1999	3.7	7.7	11.4
2000	4.0	8.0	12.0

Figures based in part on forecast prepared by Logica.

The Cable News Network (CNN)

Today there are several international news-gathering agencies. A new generation of broadcasters has appeared, mainly because of developments in portable Ku-band earth stations coupled with the developments in portable news-gathering equipment of which the most important item is the ENG camera.

Of the BBC, ABC, NBC, CBS, ITN, WTN, Visnews, NPR, AP, UPI and CNN, without doubt it is the last named, Cable News Network, which has pioneered the way. The employment of satellites for the relaying of television programmes had its origins in the USA when the two American broadcasting giants NBC and CBS relayed television programmes from coast to coast.

CNN began life 20 years ago when Robert Edward Turner made his prophecy that satellite communications and dissemination of news shared a common destiny. Today CNN has become the world standard against which all other news agencies are judged. Turner must be credited with seeing ahead of others that television broadcasting, satellite communications and international news gathering were all undergoing some remarkable, and even revolutionary, changes; news had become a great force in the world.

The other ingredient in CNN's success is its adherence to neutrality and objectivity. Ten years ago it reached out to 1.7 million homes in the USA. Today it reaches out to over 56 million homes in the USA alone and to countless millions outside that country. CNN broadcasts news at a glance, detailed news reporting, and covers international events wherever the news may be breaking. CNN broadcasts to 102 countries and territories (Figure 14.4). There is little doubt that the successful operations of CNN have spurred on broadcasting agencies like the BBC and Deutsche Welle into starting television broadcasting by satellites.

CNN operates from its international headquarters in Atlanta, Georgia, where it receives international news via satellite at its earth station where it is re-packaged. CNN Operations then feed these programmes out to transponders

Table 14.3 *The Eutelsat satellites*

	I F1	*I F2*	*I F4*	*I F5*	*II F1*	*II F2*	*II F3*	*II F4*
Stabilization	Three axis	Three axis	Three axis	Three axis	Three axis	Three axis	Three axis	Three axis
Mass at launch (kg)	1048	1172	1172	1172	1870	1870	1870	1870
Mass in orbit (kg)	512	550	550	550	915	915	915	915
Wingspan in orbit (m)	13.8	13.8	13.8	13.8	23.6	23.6	23.6	23.6
Electrical power (W)	1100	1100	1100	1100	3000	3000	3000	3000
Lifetime span (years)	7	7	7	7	7	7	7	7
Frequency bands (GHz)	14/11	14/11/12	14/11/12	14/11/12	14/11/12	14/11/12	14/11/12	14/11/12
Number of transponders	12	14	14	14	16	16	16	16
Number in use (max.)	10	10	10	10	16	16	16	16
Transmit power per transponder (W)	20	20	20	20	50	50	50	50
Antennas								
receive/transmit	–	1	1	1	1	1	1	1
receive only	2	1	1	1	1	1	1	1
transmit only	4	4	4	4	1	1	1	1
Launch date	1983	1984	1987	1988	1989	1990	1991	1991

in satellites over the Atlantic Ocean, the Pacific Ocean and the Indian Ocean. Its main satellite link with Europe is by Intelsat VI F4, with characteristics as shown in Table 14.4.

The footprint of Intelsat VI F4 enables CNN to cover the whole of Europe extending from the north, Scandinavia, to the west of Ireland, to the region of Greece in the east, and to the northernmost parts of North Africa in the south. For other parts of the world CNN operates with a variety of satellites. These are as shown in Table 14.5.

One of CNN's prime outlets is its 24-hour television news service, available in almost all the major hotel chains throughout the world.

Table 14.4

Satellite	Intelsat VI F4 transponders 13–73
Location	Nominally 27.5 W
Beam	East spot beam with beam centre on Lyon
Nominal EIRP	47.5 dBi at beam centre
Downlink frequency	11 155 MHz
Polarization	Linear
Bandwidth	30 MHz
Modulation	FM

Table 14.5

Satellite	*Operator*	*Region of the world*
Intelsat V F8 180E	Comsat	Australia, Far East and Indonesia
Galaxy 1 134W		North and Central America
PanAmSat 1 45W	PanAmSat	Latin America, Spanish-speaking countries
Stationar 12 40E	CIS	India, Middle East and parts of the CIS
Intelsat VA F12 359E	AFRTS net	These satellites cover the Far East, South America, Africa and certain parts of the USA
Satcom F2R	RCA	
Intelsat V F8 180E	AFRTS net	

The role of the satellite in the Gulf War

This was a war in which technology played the lead role. Nowhere was this more evident than to the millions of viewers in America, Europe and elsewhere in the world seated in front of their television receivers. For the first time in history they could see a war in progress, in real time direct from the theatre of operations and also from the city of Baghdad. It was truly the first video war and it was all made possible by satellite communications.

Some set up their own portable Ku-band uplinks mounted outside the El Rashid hotel in Baghdad, reporting events as they actually happened. The Gulf War was instrumental in the bringing together of many facets of video technology, ranging from video recording formats to satellite earth stations varying in size from the small 1.8 m dish to medium-sized earth stations mounted on 4 × 4 trucks.

The major satellite communications networks co-operated on a scale without precedent in order to get live pictures from the war zone into viewers' homes. Video signals from Baghdad could be uplifted to an Arabsat satellite, and from here they were downlinked to Amman where they were again uplinked to an Intelsat satellite for transmission to the east coast of the USA. To be assured of access to transponders the news agencies bought up air time on a number of satellite companies so as to be sure of bridging the distance from the Gulf to the east coast of the USA, a distance of 4065 air miles.

The electronic news-gathering (ENG) crews in the allied camp had the benefit of using the very powerful earth stations in Dhahran and Riyadh to uplift their pictures to at least two satellites. Figure 14.5 and Table 14.6 show how at least one news agency managed to get its pictures back to London and the USA.

One of the problems encountered was look angles. Because of the distances to the satellite bird, some portable uplift stations operated with look angles down to 7 degrees of elevation.

Table 14.6

Circuit Baghdad–USA

Uplink to Arabsat satellite
Downlink to Amman in Jordan
Uplink to Intelsat AOR 338
Downlink to Holmdel NJ
Uplink to Galaxy VI satellite
Downlink to Atlanta, Georgia

Circuit Baghdad–London

Uplink to Intelsat IOR 57
Downlink to London
(This same route was used by Riyadh)

Circuit Dhahran–USA

Uplink to Intelsat AOR 338
Downlink to Holmdel NJ
Then relayed onto Atlanta, Georgia
via domestic satellite Galaxy VI

Training by satellite

Distance learning, as the name implies, permits students and others to learn by means of a delivery system. The first was, of course, the postal delivery, from which was derived the name of correspondence courses. Since the early days a number of other delivery systems have come into play, the newest to arrive on the scene being training by satellite. In recent years this method has gained a foothold, though the number of people using this method is still relatively small. One of the first countries to harness twentieth-century

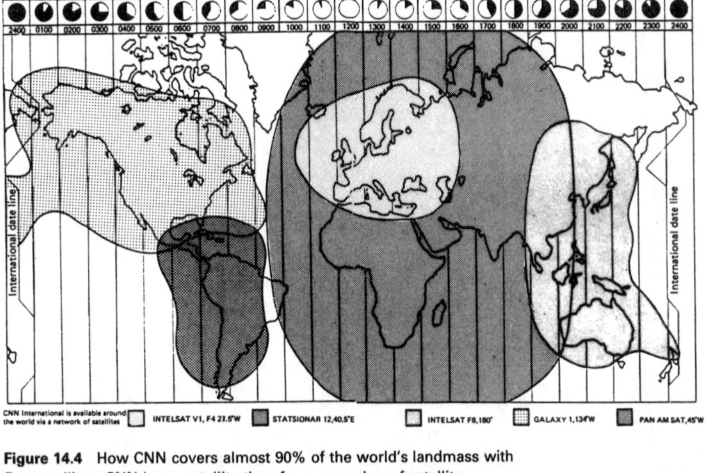

Figure 14.4 How CNN covers almost 90% of the world's landmass with five satellites. CNN lease satellite time from a number of satellite operators (Copyright 1991 by Cable News Network. All rights reserved)

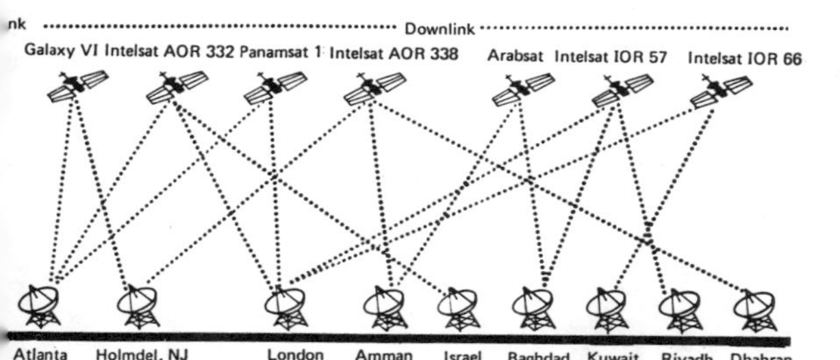

Figure 14.5 One network's SNG feed paths and satellites used to bring the story back to the USA (Courtesy of Intertec Publishing Corp.)

technology to training was the USA, using line-of-sight repeater stations in an MMDS system.

In Europe and in Asia it is likely to be satellites that will play the important role in the future. One of the main advantages of satellite delivery is the potential to reach hundreds of thousands of people with a single transmission, an element already familiar to providers of commercial satellite television. The use of DBS for teaching purposes has already been demonstrated in China, India and other countries with large landmasses and large populations but with a chronic shortage of teachers and universities.

Not surprisingly, it is the commercial operators of satellite who are casting eager eyes on this new market which is likely to gather momentum in the next five years.

Brightstar

One of the newest satellite news transmission agencies is Brightstar. Brightstar is a subsidiary of Reuters International along with VisNews which is the news gathering arm of Reuters. Reuters, the world's first news agency, began its history by flying pigeons from Achen to Paris with news messages in 1840, so Reuters' expansion into satellite communications is a prime example of how the pioneer companies keep in step with every evolutionary development in communications.

Like most other satellite news transmission agencies Brightstar does not own satellites but buys transponder time on certain satellites for the purpose of offering to its customers speedy facilities for the transmission of news (Figure 14.6). Brightstar made history in a modest way by being the first to sign up for transponder time on the Intelsat K satellite.

The crucial ingredient for providing satellite transmission services is the teleport terminal. Brightstar operates three powerful earth terminals in Washington DC, London and Moscow, and through these Brightstar channels its traffic over East-to-West, and West-to-East satellite transatlantic paths. In

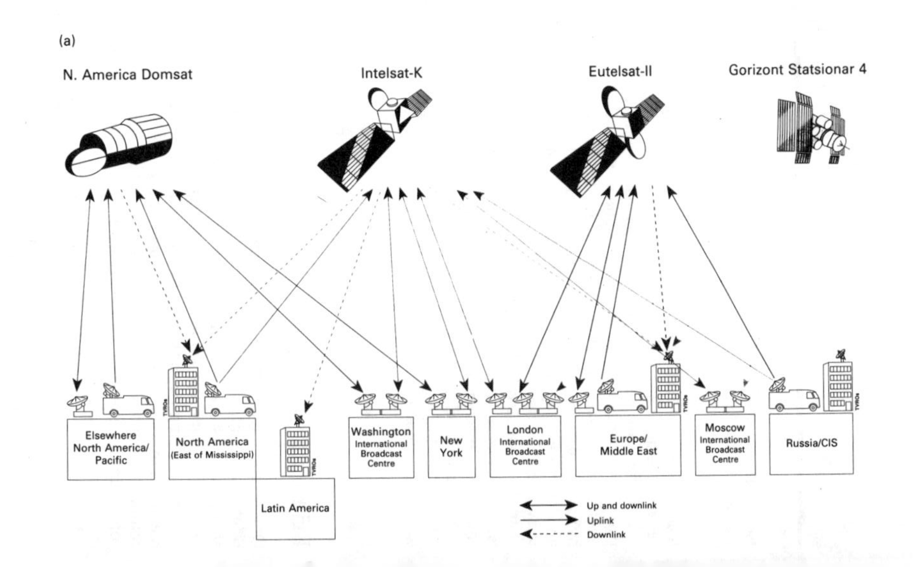
(a)
N. America Domsat
Intelsat-K
Eutelsat-II
Gorizont Statsionar 4
Elsewhere North America/ Pacific
North America (East of Mississippi)
Latin America
Washington International Broadcast Centre
New York
London International Broadcast Centre
Europe/ Middle East
Moscow International Broadcast Centre
Russia/CIS
TVROs
Up and downlink
Uplink
Downlink

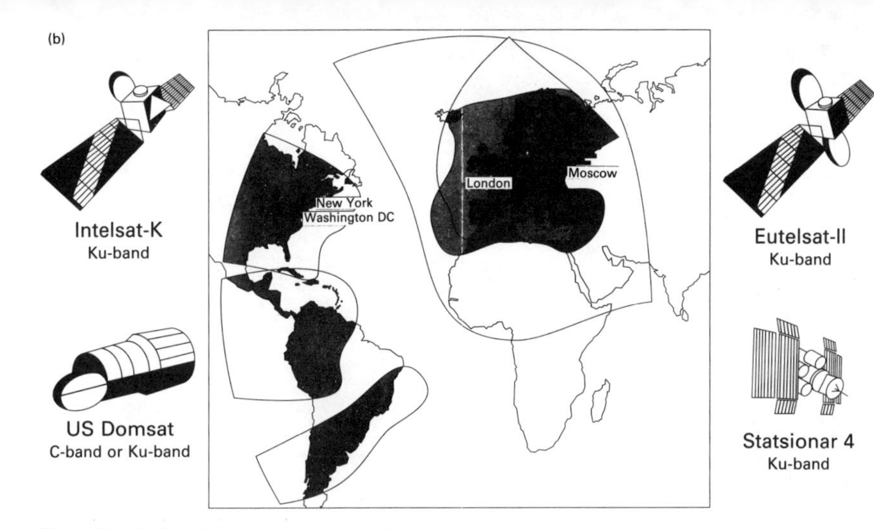

Figure 14.6 Brightstar interconnecting networks (Courtesy of Brightstar)

addition it operates a route Moscow–London and also a route from North America to Latin America through US domsats (Table 14.7).

Brightstar also operates SNG vehicles and some TVRO stations permitting SNG from as far apart as Scandinavia, the Persian Gulf and Latin America.

Table 14.7

Transmission path	*Via satellite*	*Band*
Moscow–London	Gorizont Statsionar 4	Ku band
Europe–Europe	Eutelsat II	Ku band
Europe–N America	Intelsat K	Ku band
N America–Latin America	Various US Domsats	KU/C band

15 VSATs and earth terminals

The VSAT age began on 20 November 1975 when the *Wall Street Journal* made the first ever private data transmission using a communications satellite.

Such has been the growth explosion in VSATs that today in the United States alone there are over 25 000 terminals in service, with more potential users moving into the practice of using satellites for business purposes. There is no precise definition of a VSAT; the term stands for Very Small Aperture Terminals and relates mainly to business users. As a result of developments in related technologies (large scale integrated circuits, satellites and micro computers) the modern VSAT has been much reduced in size.

Whilst size and cost have gone down, performance has gone up. Today a high performance VSAT fully interactive terminal can be bought for as little as 5000 US dollars. The VSAT market has rapidly expanded in the U.S. to take in departmental stores, small stores, gasoline stations along of course with users like railways and airports.

VSATs can be configured in a number of different ways as a system; the broadcast network, total node connection (Figure 15.1) or the star mode (Figure 15.2). However all these can be broadly divided into two categories:

(a) Fully interactive terminals
(b) Broadcast networks

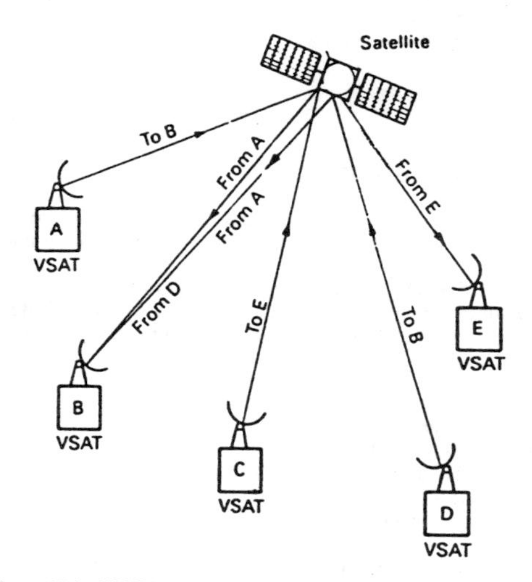

Figure 15.1 VSAT system with total node connection (*Source:* B. Ackroyd, *World Satellite Communications*)

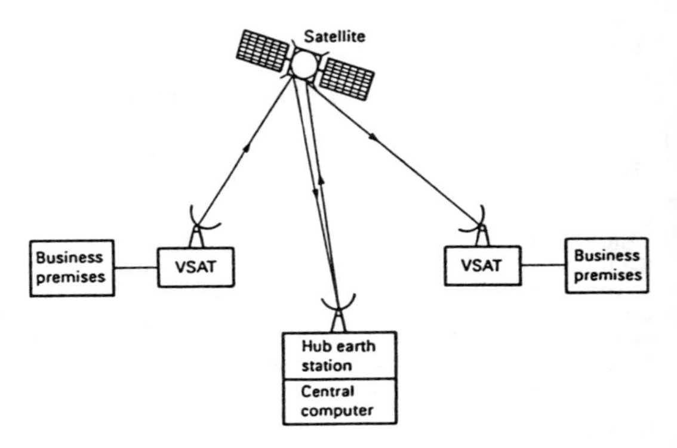

Figure 15.2 Typical VSAT star network (*Source:* B. Ackroyd, *World Satellite Communications*)

The first VSATs operated in C band, but with developments in satellite communication technology nearly all present day VSATs operate in the Ku-band. The availability of more satellites brought about a lowering of transponder leasing rates making VSATs economically attractive. Advantages of VSATs are:

(a) Low in cost
(b) Simple to operate
(c) Reliable communciations
(d) High availability rate
(e) Can be small and compact

VSAT applications

Considering the simplest form of VSAT, with no uplink to the satellite, the VSAT is simply a receiving installation to an end terminal.

An interactive VSAT, that is one which works in a duplex mode, is more costly because it has an uplink to the satellite. It also has considerably more applications to business users because it can provide the following facilities:

(a) Computer to computer interchange of data
(b) Video conferencing
(c) Point of sale facility
(d) Airline/rail network/airline ticket reservations and confirmations
(e) Educational systems, e.g., distance learning

Fully interactive VSATs are more complicated, needing a powerful hub earth station. This is crucial to the system: the gain on the uplink from the HPA and the gain on the receive

Table 15.1 *Standard earth stations*

Type	*Antenna size*	*Service*	*Frequency band*	
A	15–18 m	International voice, data & TV, IBS and IDR	6/4	C band
B	10–13 m	International voice, data & TV, IBS and IDR	6/4	C band
C	11–14 m	International voice, data & TV, IBS and IDR	14/11	Ku-band
D1	4.5–6 m	Vista	6/4	C band
D2	11 m	Vista	6/4	C band
E1	3.5–4.5 m	IBS	14/11, 14/12	Ku-band
E2	5–5.7 m	IBS and IDR	14/11, 14/12	Ku-band
E3	8–10 m	IBS and IDR	14/11, 14/12	Ku-band
F1	4.5–5 m	IBS and IDR	6/4	C-band
F2	5.5–7 m	IBS and IDR	6/4	C band
F3	9–10 m	International voice and data, IBS and IDR	6/4	C band
G	All sizes	International lease services ILS	6/4 14/11, 14/12	C and Ku-band
Z	All sizes	Domestic lease services DLS	6/4 14/11, 14/12	C and Ku-band

IDR = International data rate; IBS = International business services; ILS = International lease services; DLS = Domestic lease services; Vista = low capacity earth stations (Courtesy of Intelsat) leg down to subscriber terminals supply the overall system gain to counteract free space signal loss and thereby permit the use of VSATs with small to medium sized dishes.

Almost all of the manufacturers are US companies. Market leaders with a specialist capability in VSAT technology include:

(a) Hughes Network Systems
(b) GTE Spacenet
(c) AT&T Trydom
(d) Intelesys
(e) Radiation Systems
(f) Scientific Atlanta
(g) ViaSat Technology Corp

The design of a VSAT system has to take account of the requirements, i.e. broadcast or interactive and type of traffic (data/telephony/fax). Encryption may be a requirement and particular attention has to be given to computer interfacing, data rate, error rate, and signaling protocol.

A typical specification for an interactive, portable VSAT:

(a) Highly portable and easily set up by one person
(b) Complete with flyaway 1.2, 1.8 or 2.4 metre dish
(c) Suitable for both C band and Ku-band operation
(d) Built-in inclinometer, compass, and polarisation device
(e) Full duplex operation, and 64 kbit/s channel capable of providing 8 voice and 4 data channels
(f) AC or DC powered, with low consumption.

The basic form of any earth station is the same regardless of the system in which it is used. The elements of the system are shown in Figure 15.3 and can be regarded as a two-way microwave communication link that requires certain specialized elements in order to operate in the particular environment associated with satellite communications.

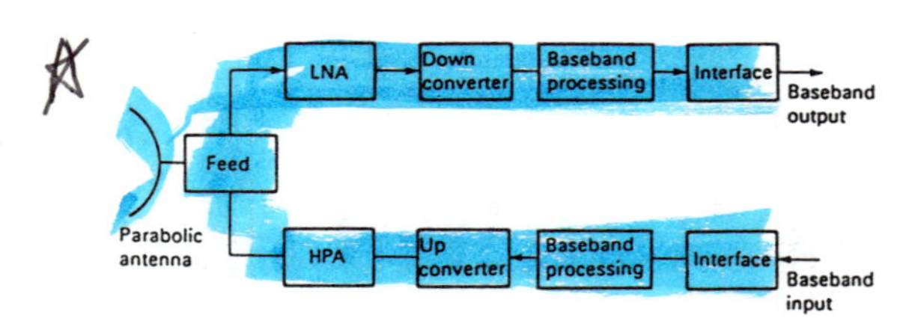

Figure 15.3 Basic earth station

16 Mobile satellite systems (MSS)

The age of mobile satellite services (MSS) was inaugurated in 1979 with the birth of Inmarsat. Its purpose was to provide a reliable world-wide maritime communication system for ship-ship, and ship-shore. Inmarsat is an internationally owned consortium with 67 member nations.

Inmarsat's main role is in maritime communications and with the aid of satellite communications it has set performance standards that never would have been possible with HF communications. As at December 1991 there was a total of 12 875 Inmarsat Ship Earth stations in maritime use, an increase of almost 20% over the previous year. The standard A/B radome and terminal is now a familiar sight on all medium to large vessels. These terminals provide direct dialling, fax, data, and telex on a 24 hour basis. In January 1991 Inmarsat began its second maritime service intended for much smaller ships, the Inmarsat C service. Inmarsat C is similar to the A/B larger vessel service but with lower rate of service. It is intended for coastal and shipping vessels.

In the Inmarsat satellite services the Land Earth station (LES) provides the crucial link between the satellite and the appropriate national telecom network. The number of LES stations operating around the world as at December 1991 was 24 for the Inmarsat A Service and 8 for the Inmarsat C Service. The LES stations replace the coastal stations of the wireless age and are still referred to as coastal stations.

In 1991 Inmarsat moved into the business of providing aeronautical services. Aviation services are in three categories:

1. A store and forward service for aircraft in all parts of the world
2. A low gain real-time service for communications between airports and aircraft
3. Inmarsat high gain services providing multiple channel flight deck voice and passenger telephony services.

Aeronautical services are a part of Inmarsat's business which are expected to show rapid growth.

In 1992 Inmarsat inaugurated its Land Mobile Service called Inmarsat M. It uses the existing satellites as used for the other services, the user's terminal is housed in a briefcase and it supports telephony, group 3 fax, and duplex data.

Looking further ahead Inmarsat will introduce a paging system in 1994, and by possibly the year 2000 Inmarsat will have inaugurated the world's first global, portable hand-held telephone. However the concept of a global portable telephone system enabling a person to talk to the other side of the world with a small hand-held portable is attracting attention from many sources.

Most of the proposals are revolutionary in the sense that they have moved away from the current techniques of employing geostationary satellites, and towards the employment of smaller low-earth orbiting satellites (LEO). At the present time Inmarsat is examining all options of satellite configuration (Figure 16.1):

Option 1 The existing geostationary orbit
Option 2 A combination of LEO and GSO satellites
Option 3 A rather more complicated pattern of intermediate circular and geostationary orbit.

Satellite system

To achieve the coverage, capacity and features required for hand-held satellite telephone services, new satellites will be needed. A number of different proposals are currently being considered by Inmarsat, including an examination of non-geostationary orbits.

Among the options under evaluation are:

(a) an enhanced geostationary earth orbit (GEO) satellite system, similar in orbital configuration to but more powerful and with larger antennas than Inmarsat's existing satellites
(b) a low earth orbit (LEO) satellite system overlay
(c) an intermediate circular orbit (ICO) satellite system overlay
(d) a combination of GEO and non-GEO constellations, with inter-satellite links between the non-GEO and GEO satellites.

The satellites

Inmarsat uses its own Inmarsat-2 satellites, leases the Marecs B2 satellite from the European Space Agency, maritime communications subsystems (MCS) on several INTELSAT V satellites from the International Telecommunications Satellite Organization and capacity on three MARISAT satellites from COMSAT General of the United States. The system is currently configured as shown in Table 16.2.

Each Inmarsat-2 spacecraft, the fourth and last of which was launched in April 1992, has a capacity equivalent to 250 Inmarsat-A voice circuits.

Inmarsat has contracted with GE-Astro for an Inmarsat-3 series of four larger satellites for launch beginning in 1994 (Table 16.3) and is now investigating a number of options – including non-geostationary – for satellite systems for the twenty-first century under its Project 21 initiative.

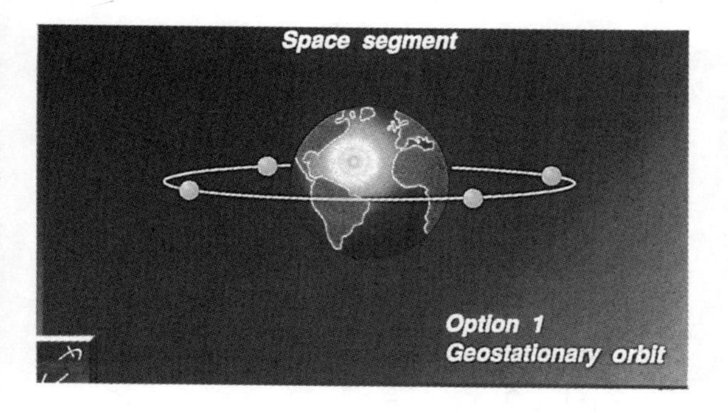

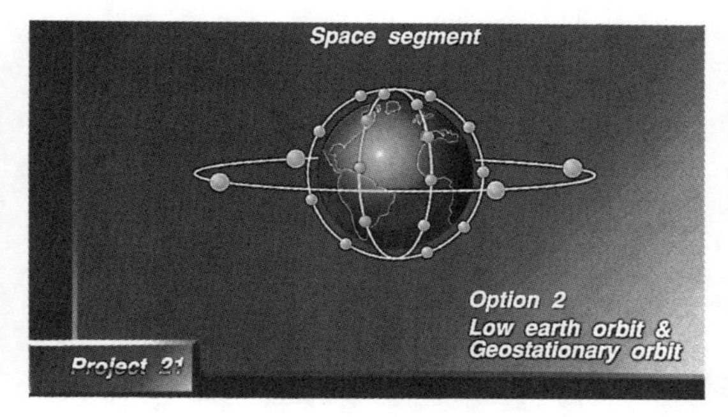

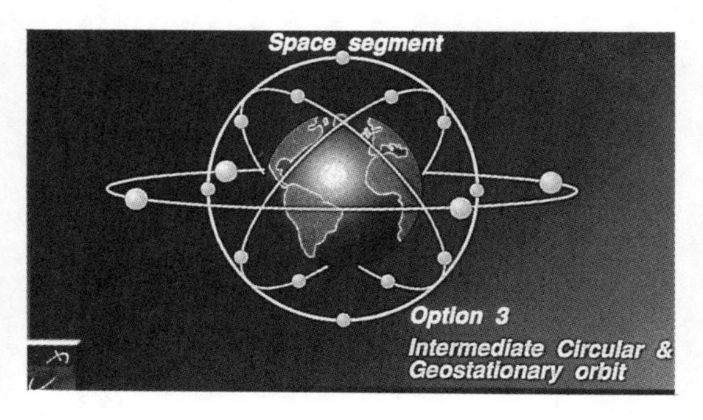

Figure 16.1 The Inmarsat evolution (Courtesy of Inmarsat)

Table 16.1 *A growing family of services*

Service	*Features*	*Remarks*	*Markets*	*Date*
Inmarsat-A	HSD, F, Tel, Tx	Full service, high quality terminal	LT, M	Introduction 1982
Inmarsat-C	LSD, Tx	Low-cost portable	LM, M	Introduction 1990
Inmarsat Aero-L	LSD	Commercial, business and private aircraft	A	Introduction 1990
Inmarsat Aero-H	F, HSD, T	Commercial and larger business aircraft	A	Introduction 1990
Inmarsat-B	F, HSD, T, Tx	Full service, high quality digital terminal	LT, M	Introduction 1992
Inmarsat-M	F, HSD, T	Medium quality, lower cost digital portable	LM, M, P	Introduction 1992
Inmarsat-E	F, HSD, T, Tx	Instant maritime distress alterting system	M	Introduction 1992
Inmarsat Paging	LSD	Pocket-sized, global paging system	P	Introduction 1994
Inmarsat Navigation Services	F, HSD, T, Tx	A range of services for navigation purposes	A, LM, M, P	Introduction mid 1990s
Inmarsat-P	F, T	Hand-held, low-cost global satellite telephone	P	Possible introduction by 2000

Aeronautical = A; Facsimile = F; High speed data = HSD; Land mobile = LM; Low speed data = LSD; Land transportable = LT; Maritime = M; Personal = P; Telephone = T; Telex = Tx. (*Source: Inmarsat Annual Review, 1991*)

Table 16.2

Ocean Region	Atlantic(W)	Atlantic(E)	Indian	Pacific
Operational Location:	Inmarsat-2 F4 55W	Inmarsat-2 F2, Marecs B2 15.5W 15.2W	Inmarsat-2 F1 64.5E	Inmarsat-2 F3 178E
Spare Location:	Intelsat MCS-B 50W	Intelsat MCS-A Marisat F2 66E 72.5E	Intelsat MCS-D Marisat F1 180E 106W	Marisat F3 182E

Table 16.3 *Future satellites and launch dates*

Sat manufacturer	*Type*	*Launch contracter*	*Vehicle*	*Year*
GE Astro	Series 5000	General Dynamics	Atlas	1994
"	"	"	"	1994
"	"	Arianaspace	Ariana 4	1995
"	"	Russia	Proton	1995

Table 16.4 *Specification for a typical lightweight transportable communication set*

Requirement: hand carried high quality fax, data and voice capable of contacting any telephone number world-wide.

Size:	61 × 27.3 × 35.5 cm
Weight:	19 kg
Dish:	1.2 metres parabolic
Dish gain:	23.5 dBi
Power requirement:	100–375 VDC, 47–440 Hz, 180 watts
Transmit:	70 watts idle
Environmental:	−35 to + 55°C RH 95% max at 40 °C
Precipitation:	10 cm/hour max
Frequency:	transmit between 1636.5 and 1645.0 MHz receive between 1535.0 and 1543.5 MHz
Telegraph channel:	Modulation BPSK, data 72 bits (12 characters/6 bits)
Voice channel:	Modulation FM 300–3000 Hz
Carrier EIRP:	36 dBW +1, −2 dB

Note: Setting up is in four easy steps: open suitcase; pop open the umbrella antenna and connect; plug into power source; place the call.

This specification relates to one model produced by Mobile Telesystems Inc., fully type approved by Inmarsat. (Courtesy of MTI Mobile Telesystems Inc.)

Global communications from hand-held portable telephones

A revolution in communications technology is now in progress and by the year 2000 it is almost certain that the cellular portable communications will be supplemented with a global based network, using satellites with much less power than the giant geostationary which at present are the mainstay of all communication systems whether FSS, BSS or MSS.

Inmarsat regards global mobile communications as being within its scope. To this end Inmarsat has commissioned Project 21 which is a study programme aimed at developing a revolutionary global communications system for mobile use. Various options will be considered ranging from geostationary to LEO or a combination of both.

This said, it may emerge that one of the most important developments of the twenty-first century will be global communications enabling a portable phone to call another subscriber anywhere in the world, at any time of day or night with the same degree of reliability. Some of the proposals are outlined in Table 16.5.

Table 16.5

Project	*Satellite configuration*
Iridium	66 low power LEOs
Odyssey	9 low power LEOs
Orbcom	20 small sats, 2 polar orbiting and 18 equally spaced in three planes, inclined at 40° to the Earth
Project 21	Not yet determined
Globalstar	24 low power LEOs

(Courtesy of Mobile Telesystems Inc.)

17 Emergent technologies

The technology of fibre optics

It is hard to predict a long-term future for satellites in view of the potential of fibre optics and the speed of progress in this new branch of communication sciences. Fifteen years ago the term 'fibre optics' was new. Though a use for this technology was forecast by many, few at that time would have even dared to be so foolish as to predict it as a global means of communications. Yet by the year 2010 it may be the only link between home and office; it will be the link with the outside world, carrying radio and television, and it will link the home with the library and a whole host of other facilities, including mail order. Fibre optics offer the first real prospect for an integrated services digital network (ISDN).

The idea of using light, modulated in such a manner as to transmit intelligible speech, is not new. The suggestion had been put forward by scientists in the mid-1850s, but the problem was that no one knew how to accomplish the task; in short, it was a good idea but the technology simply did not exist. The invention of the laser in the early 1960s gave a new impetus to the idea. This was because the laser gave the designer a light source that could be modulated. By the late 1960s it had been discovered that light could be transmitted along glass fibres considerably more easily than along conventional cables.

The first commercially manufactured fibre optic cables came into being by the early 1970s and heralded the age of optical communications, what might be described as the third wire age. It was soon discovered that fibre optic cables possessed tremendous possibilities; perhaps the most important features were a cable attenuation loss less than one tenth that of conventional cables, and a bandwidth almost a hundred times greater.

Fibre optic cables work on the principle that if light is transmitted along an optical fibre then the use of a central core with a higher refractive index than the outer cladding will ensure total internal reflection.

The main advantages of fibre optic cables over conventional cables, coaxial or otherwise, are:

(a) extremely low attenuation
(b) total immunity from electrical interference
(c) wide bandwidth (up to 1000 MHz)

On the other hand, with the present state of the art, the handling, laying and terminating of fibre optic cables requires some expertise.

In terms of signal purity, fibre is unmatched. The greater the distance, the more fibre optics comes into its own, and for transcontinental applications it is unmatched by any other form of communication. Signals transmitted over fibre optics are:

(a) not affected by fade-outs
(b) not affected by magnetic storms
(c) not affected by ionospheric conditions
(d) not affected by local electrical interference
(e) not affected by attenuation over long distances
(f) not affected by rainstorms or snowstorms, as is microwave.

Telecommunications corporations are not the only ones who are watching the advances in fibre optics. Increasingly, the cable television industry in America is getting interested. Until now cable television has relied upon coaxial cable, but now all this is changing. At one time the broadcasting companies thought that there was no alternative to cramming an HDTV signal into a 6 MHz bandwidth so as to make the HDTV signal suitable for transmitting down a coaxial cable; now there is no need if fibre optic cables replace coaxial.

There are other advantages in making the switch to fibre optic from coaxial. Coaxial systems need several amplifiers on a long run. Every amplifier, modulator and demodulator is a source of further signal degradation. A present-day large-cable satellite delivery system might have as many as 50 amplifiers between the head-end plant and the end-subscriber's home. Moreover, because fibre optic systems transmit light instead of electrical RF signals the overall reliability of the system must be considerably improved.

Fibre optics have their shortcomings on very short runs. Fibre optics are most costly to install, calling for a higher degree of skill in making connections. Fibre optic cables do not like going round corners but all of these disadvantages will gradually become of less significance as more experience is gained on local systems. In the short term, cable television companies will transpose from fibre to copper at the consumer premises.

It has been estimated that the 25 countries with the highest telecommunications expenditures are expected to spend some 375 billion US dollars between now and the next four years on investment in new technologies, and the main item of expenditure will be on fibre optic systems. The US Department of Commerce has estimated the value of the fibre optics systems market worldwide at 3.8 billion dollars for 1990, with an expectation of growth by the year 2000 to 10.8 billion dollars. The US portion of this market was 1.4 billion US dollars in 1990 and is expected to grow to 4.6 billion dollars by the end of the decade. Elsewhere in the world the fibre optics market is expected to at least double.

Global economic wealth is concentrated in three regions of the world: North America, Asia/Pacific and Western Europe. It is from these three regions of the world, and especially Japan, that the greatest growth will come. By the year 2000 we are likely to see fibre optic super highways linking these continents together for the transmission of voice, data, and video, i.e. HDTV.

Operators of fibre networks

Those who are watching the future of fibre very carefully are the telephone companies. Just as the telephone companies of 70 years ago were the only people with the means of relaying a radio broadcast over a few thousand miles, so the very same telephone companies are making sure they will control the successor to the metallic pair of wires. At the present time for some telephone companies the problem is financing the laying of a national fibre optic network, but come it must. The potential for fibre optics is so immense that it is possible to conceive a future, perhaps in less than 30 years, when the definition quality of HDTV will be superseded by new definition standards of super-HDTV which will make 1125/60 and 1250/50 seem like the early 1936 days of television broadcasting.

The future concept of a digitally integrated broadband delivery network for the delivery of perhaps hundreds of channels of voice and data, to say nothing of delivery of television, has sparked off interest by governments and commercial companies. The costs are so great that no one is able to calculate exactly what it is going to cost, or how to go about pricing the data transmission rates. What is certain is that present-day methods of costing transmission time will not be acceptable for the future. For example, if the telephone company prices rates at, say, one penny per minute for a 64 kbit/s service (voice transmission), then a two-hour movie delivered at 45 Mbit/s would cost over 800 US dollars.

Perhaps more so than in any other part of the world, the US manufacturers are leading the way in preparation for the day when fibre optic networks will have taken over from traditional coaxial cable. One of the leading companies in this technology is Jerrold Communications, a subsidiary company of General Instruments. Jerrold Communications has developed terminals to provide optic links up to 19 km. With a performance of 55 dB carrier to noise, and 65 dB carrier distortion, fibre optic links will transform cable television.

Digital video compression

Digital video compression is the newest technology to arrive on the scene and it promises to revolutionize television broadcasting and satellite communications. Although the digital revolution began only a few years ago it is rapidly gathering momentum. Its implementation has already begun in certain sectors, notably business video systems and also in the sector of distance learning; now the cable companies and the television broadcasting industry including the DBS sector are about to jump on the waggon.

Although DVC is a new technology, audio compression techniques have been with us for many years. In essence audio compression is to put as much audio as possible into a specified bandwidth whereas video compression is to reduce the bandwidth by as much as possible without significantly impairing the quality of signal. In video compression sometimes the term bit data reduction is used, at other times the word compression is used. Both amount to the same thing: the objective is to reduce large amounts of visual and audio data to smaller, more manageable levels without significant and obvious loss of quality.

The greater the degree of compression the more of the original data is taken away. In the long term it is possible that telecommunications will be using compression rates of 100 : 1. There are numerous applications where high compression rates can be employed but these do not include TV and HDTV broadcasting. However the immediate objective of the DVC lobby is to use DVC for the purpose of compressing an HDTV signal, presently needing a 27 MHz bandwidth, down to a 6 MHz bandwidth thus enabling HDTV to be re-transmitted over UHF-TV transmitters. Eperimmmmeental over-the-air tests have demonstrated that a compression rate of 4 : 1 is possible without any detriment to the transmitted HDTV signal.

There are significant advantages to be gained from broadcasters going over to DVC technology, the most important being a cost-saving on transponder. For example four HDTV signals presently needing a 27 MHz bandwidth for each, could now be digitally compressed so that all four TV programmes could be radiated from one transponder. Against is the high cost of going over to employing DVC. With the present state of the art compression equipment is costly. At the TV transmitter the programme must be digitally

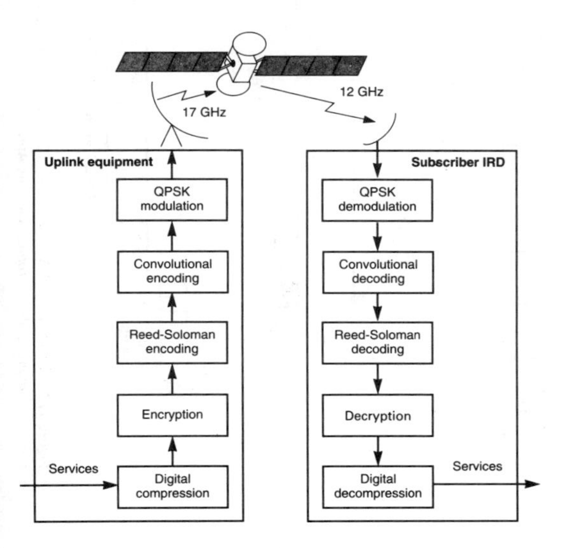

Figure 17.1 Block diagram of DirecTv DBS system (*Source:* L. Lockwood, DirecTv, a digital DBS, *International Cable*, April 1993)

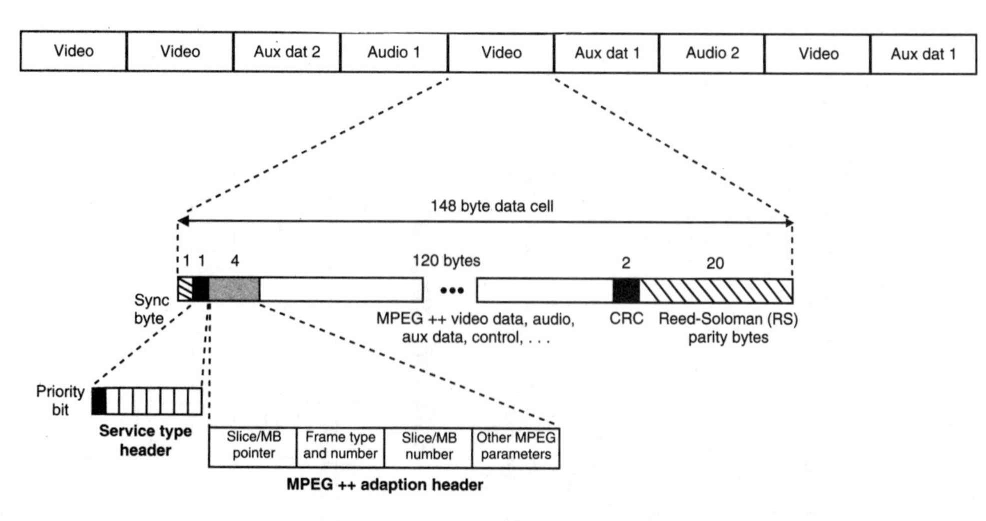

Figure 17.2 Overall AD-HDTV data system (*Source:* L. Lockwood, DirecTV, a digital DBS, *International Cable*, April 1993)

compressed, and it follows that every TV receiver must have equipment to expand the TV signal to its original format.

For these reasons the first to employ digital compression will be the cable companies because it is only at the head end where DVC expansion of the signal is needed. Thereafter the TV programme is distributed by cable systems to its viewers.

If the television programme transmitted is by analogue delivery, then an analogue to digital (A/D) conversion must be used before the video digital compression can take place. This is why America will adopt an all-digital solution for its HDTV, and why ultimately Japan and Europe must follow suit. Eventually, DVC technology will become cheap enough to incorporate into HDTV receivers without adding a great deal to the cost.

At the present time there are a number of companies developing digital video compression systems; Scientific Atlanta with its MPEG-2 system, and the joint venture agreement between AT&T, and GI, with its DigiCipher System. Another company in the vanguard is the US company CLI; its advanced CDV technology called SpectrumSaver digitizes and compresses a PAL or NTSC television signal so that it can be transmitted in as little as a 2 MHz bandwidth.

Compression technology and definitions

For a number of years two expert groups have been studying video compression. These are the Joint Photographic Expert Group (JPEG) along with the Motion Picture Expert Group (MPEG). Video compression is a highly complex science involving perception and reality. Developments are proceeding at a fast rate. Each new breakthrough reveals that we may have merely scratched the surface of a new technology. Aural and visual psychophysics are just two of the studies involved with it. At the present time the techniques used may be defined:

(a) *Compression* The compression of data rates by the employment of statistical high order mathematics to remove redundant information. What is discarded is determined by relative dependencies, motion predictions and relative entropy levels.

(b) *Bit rate reduction* Reduces data by discarding information superfluous or imperceptible under static viewing conditions.

(c) *Perception-based bit reduction* Works by discarding information that cannot be seen or heard. The human brain makes use of adaptive information selection by deciding what it needs to see or what to hear.

(d) *Algorithm* An instruction set that defines how the input data is dis-assembled, compressed or truncated and coded. At the receiving end another algorithm is needed to reconstruct the waveform.

Most of the data reduction systems used use one of three technologies: digital compression; analog compression; or perception-based bit rate reduction.

Digital audio broadcasting (DAB)

Digital radio is developing in fits and starts, and moving in different directions at the same time. The divergence between one country and another takes in different formats and different frequencies. For Europe the Eureka 147 is the favourite and the World Administrative Radio Conference (WARC 92) held in Torremolinos, Spain defined the world standard at L Band 1452–1492 MHz.

Deviations from this standard are being made in the USA, CIS, Japan and others. At the time of writing this book insufficient data or experience is known about DAB except that we are told it will be a giant leap forward in quality of transmission over existing radio services carried out on AM and VHF-FM wave bands. It is also uncertain as to what level of transmission power will be needed. One estimate based purely on theoretical considerations put the required transmitting power as 4 times that used for VHF-FM to realize the same area of coverage.

If the transmission position is not all that clear the same applies to the radio receivers. No one has any clear ideas on what these will cost, but to ensure success then the cost should not exceed that of a good quality FM receiver. DAB, when it comes, will be broadcast over terrestrial and satellites.

The uncertainty of when DAB will reach the man in the street is best illustrated by the fact that in the USA alone there are eight different systems under consideration, only two of which have had a practical demonstration. There are some experts who believe less attention should be paid to emerging technologies like DAB, and more attention paid to finding solutions to the problem of multipath distortion in FM reception, and improving what we already have.

18 Conclusions

It is difficult to make any accurate prediction on when HDTV will become universal throughout the world, and the existing formats PAL, SECAM and NTSC will become history. On the one hand there is the quest for better picture quality, with initiatives coming from the manufacturers. On the other hand there is an almost insatiable appetite from the viewer for a greater choice in channel viewing; this demand is being catered for by the programme suppliers, satellite broadcasters and cable companies.

In this contest between quality and quantity it is the latter which is winning greater viewing audiences, and the competition may only be resolved when more and more viewers have seen a large screen television presenting HDTV. Aspirations must be towards a picture quality that comes as close as possible to the cinema viewing experience. At the present time none of the HDTV systems in use can produce cinema-film quality.

The ultimate HDTV system in the foreseeable future, that is 2010, will almost certainly be all-digital. Working towards this solution is the European Broadcasting Union (EBU). It is actively promoting an all-digital HDTV and to this end it is looking to the World Administrative Radio Conference (WARC) organized by the International Telecommunications Union (ITU) to administer a new band of frequencies for digital HDTV. This new band proposed by the EBU will be in the 12.7–22.0 GHz band; this will enable broadcasters to transmit worthwhile HDTV transmissions with a wider bandwidth than that possible in the Ku-band.

The 12 GHz band is steadily filling up with a variety of conventional analogue services along with HDTV analogue systems such as HD-MAC.

Between now and the year 2010 we are also likely to see some significant improvements in transmissions and receiver technologies. Already the trend in more powerful transponders is gathering momentum, thanks to developments in TWT technology. Soon, the medium-power satellites used by Astra will become outdated. An increase in transponder power from 45 to 230 W would ensure better picture quality. Similarly, we can expect some improvements in the design and performance of satellite receivers; this would permit viewers to use a slightly smaller dish size, or produce improved reception quality with existing sized dishes.

Whilst this is going on, developments in fibre optic transmission systems will also not be standing still. Proponents of fibre optics will argue that no over-the-air system, whether by UHF, MMDS, LDTV or satellite, can match the purity of fibre optics. With an immunity to fade-outs, magnetic storms, ionospheric behaviour, terrestrial noise and adjacent channel interference, fibre optics stands the best chance of taking over from some of these methods. One way or another, the next century will see some truly remarkable changes in the television we know today.

Abbreviations

Organizations and regulatory authorities

AOR Atlantic Ocean Region of Intelsat
ASEAN Association of South East Asian Nations
Aussat Australian National Satellite System Operating Company
Brazilsat Brazilian Satellite System
CATA Community Antenna Television Association
CBC Canadian Broadcasting Corporation
CBS Columbia Broadcasting System
CCITT International Telegraph and Telephone Consultative Committee
CEPT Conference Européenne des Administration des Postes et des Telécommunications
CNN Cable News Network
Comsat Communications Satellite Corporation
CONUS Continental United States
EBU European Broadcasting Union
ESA European Space Agency
FCC Federal Communications Commission
IBS Intelsat Business Services
INMarsat International Marine Satellite Organization
Insat Indian National Satellite System
IOR Indian Ocean Region of Intelsat
ITU International Telecommunications Union
NASA US National Aeronautics and Space Administration
NBC National Broadcasting Company
POR Pacific Ocean Region of Intelsat
RARC Regional Administrative Radio Conference
UN United Nations
WARC World Administrative Regulatory Committee
WMO World Meteorological Organization

Communications and satellite

ACI Adjacent channel interference
ADC Analogue to digital converters
AFC Automatic frequency control
AGC Automatic gain control
AM Amplitude modulation
CATV Community antenna television
CCI Co-channel interference
DATV Digitally assisted television
DBS Direct broadcast by satellite
DTH Direct to home (broadcasting by DBS)
ECS Electronic communication satellite
EIRP Equivalent isotropic radiated power
ELV Expendable launch vehicle
ENG Electronic news gathering
FDM Frequency division multiplex
FET Field effect transistor

FM Frequency modulation
FSS Fixed satellite service
GP Geostationary platform
GSO Geostationary orbit
HPA High-power amplifier
IPA Intermediate-power amplifier
LEO Low earth orbit
LHCP Left-hand circular polarization
LNA Low-noise amplifier
LNB Low noise block
LNC Low noise converter
LPTV Low-power television
LSI Large-scale integration
MAC Multiplexed analogue components
MCS Maritime communication system
MMDS Multipoint microwave distribution system
MSDC Multi-stage depressed collector
MSS Mobile satellite service
MUSE Multiple sub Nyquist sampling encoding
NCS Network co-ordination station
PAL Phase alternation by line encoding system
PM Phase modulation
PSTN Public switched telephone network
RA Rainfall attenuation
Radarsat Radar satellite
RF Radio frequency
RFI Radio frequency interference
RHCP Right-hand circularly polarized
RX Receiver
SBS Satellite business systems
SECAM Sequence colour and memory
SES Ships earth station
SMATV Satellite master antenna television
SNR Signal to noise ratio (or S/N)
SSPA Solid state power amplifier
TDM Time division multiplex
TDRS Tracking and data relay satellite
TTC Tracking, telemetry and control command
TVRO Television receive only
TWT Travelling wave tube
TWTA Travelling wave tube amplifier
TX Transmitter
UHF Ultra-high frequency
VCR Video cassette recorder
VLSI Very large-scale integration
VSAT Very small aperture terminals
XPD Cross polarization diversity
XY Refers to horizontal (x) or vertical (y)

Glossary

adjacent channel interference Interference between two transmissions.
amplifier A device which increases the amplitude of the signal.
analogue transmission A transmission where modulation is achieved by analogue means.
antenna Another term for aerial, a collector or radiator of electromagnetic energy.
aperture The beamwidth of an antenna.
apogee The maximum altitude of a satellite.
asynchronous operation Data sent in a stream.
attenuator A device for reducing the level of signals.
azimuth Angle of elevation between ground and antenna centre beam.
bandwidth The frequency band required for a transmission of signals.
baseband The frequency band prior to demodulating of signals.
carrier to noise Ratio of carrier level to noise level.
cassegrain Dual reflector antenna, with paraboloid and hyperboloid subreflector.
circular polarization Transmission mode in a circular pattern.
common carrier A public body able to provide a telecommunications service.
companding Signal compression and expansion. The term is derived from the first and last parts of compression and expanding.
cross-talk Where a signal interferes with another.
data channel A transmission service carrying a data stream.
decibel A unit of loudness or level based upon a logarithmic scale.
delay distortion Signal impairment due to speed of propagation.
delay time Time taken for a signal to travel to and from a satellite.
down-converter A frequency translation to an intermediate frequency.
downlink The communication path from satellite to earth station or viewer.
earth station A station which accepts programmes and transmits to satellite.
echo distortion Signal corruption due to echo combining with direct ray.
elevation The angle measured between earth and the direction of the antenna.
encryption The process of encrypting a signal to make it unintelligible without decoding.
figure of merit A measure of quality of a satellite earth station.
focal length The distance from the centre of a dish to the feed.
footprint The area or region covered by the beam of the satellite antenna.

free space loss The signal attenuation from a satellite. It is of the order of 206 dB for a satellite at 22 300 miles above the earth and at a frequency of the order of 11 GHz.

frequency re-use The use of the same frequency but with a different polarization.

geostationary satellite A satellite whose orbital speed is such that the satellite is synchronized with respect to earth.

gigahertz A unit of measurement; 1 GHz equals 10 MHz.

Gregorian antenna An antenna which uses a parabolic reflector and a concave ellipsoidal subreflector.

group delay distortion Distortion due to different frequencies being subject to different delay times.

guard band An unused portion of the spectrum separating two adjacent channels.

half-transponder operation Two television channels transmitted in a 36 MHz bandwidth.

head end The main distribution feed for cable television systems.

intermodulation The third order products resulting from the mixing of two frequencies.

jitter The distortion resulting from the advance or retard of digital signals.

kelvin Unit of absolute temperature used in noise measurement.

kilobit or k-bit Equal to 1000 binary digits.

Ku-band The spectrum between 10.7 and 18 GHz used for satellites.

link budget The summation of the factors affecting a satellite link.

low-noise amplifier The amplifier following the output from the receiving antenna.

meridian The line passing through a position to the earth's axis.

modem A modulator and demodulator built into the same unit.

modulation The process of modulating a carrier with information by either FM, AM, phase, or a combination of the three.

multiple access Where more than one user can access a transponder, by time/frequency.

multiplexer Where a number of signals are combined on a common carrier.

noise Electrical noise from whatever source.

noise factor Measure of the goodness of a system relative to zero noise.

noise temperature Measure of the noise in a system.

offset antenna A dish whose feedpoint is not on the axis of the antenna. This method of mounting reduces the amount of signal blockage.

orthogonal At right angles.

paraboloid A parabola of revolution.

phase The spacing in degrees between two waveforms on a common time axis.

polarization The direction of the electromagnetic wave.

quantization A process used in pulse amplitude modulation to transform to digital.

random noise White or gaussian noise.

redundancy As applied to satellite earth stations it means a dualling arrangement.

satellite A communications station in space.
shaped antenna reflector Shaping a reflector to modify the footprint.
sideband A method of modulation (AM) which generates an upper and lower sideband.
sidelobes The off-axis response of an antenna, i.e. unwanted responses.
solar outage Disruption of satellite communications because of radiation from the sun. It occurs when the antenna looks at the sun.
spread spectrum The spreading of modulation across a wide bandwidth.
teleconferencing A communications system that links people together by satellite.
transponder The equipment that receives the signal from the uplink to the satellite and re-transmits the same back to earth via the downlink.
uplink The signal from the earth station to the satellite.
visible arc The arc of the geostationary orbit over which the satellite is visible from earth.
voice circuit In telecommunications service it is from 300 to 3400 kHz.
white noise That due to basic movement of electrons.

Bibliography

Ackroyd, B. (1986) The Arabsat communication system, *Electronics and Power*, **32**(7)
Ackroyd, B. (1990) *World Satellite Communications and Earth Station Design*, BSP Professional Books (Blackwell Scientific), Oxford
Barnouw, E. *History of Broadcasting in the US*, Vol. 1 to 1933, Oxford University Press, New York, pp. 7–9
Barnouw, E. *History of Broadcasting in the US*, Vol. 1 to 1933, Oxford University Press, New York, pp. 19
Barnouw, E. *History of Broadcasting in the US*, Vol. 1 to 1933, Oxford University Press, New York, pp. 20
Baron, S. N. (1990) Advanced television broadcast services, *International Broadcasting Convention Technical Paper No. 327*, IEE, London, pp. 417–420
Barriere, J. M. (1988) Terrestrial transmission of D2-MAC packet, *International Broadcasting Convention, Technical Paper No. 293*, IEE, London, pp. 293–299
Bertenyi, E. (1988) Space segment of the BSB DBS satellite system, *International Broadcasting Convention Technical Paper No. 293*, IEE, London, pp. 192–195
Boegels, P. W. (1988) The Eureka Project, philosophy and practice, *International Broadcasting Convention Technical Paper No. 293*, IEE, London, pp. 430–437
Brown, G. (1990) Peoples Republic of China places restrictions on satellite dishes, *World Broadcast News*, November
Burns, R. W. (1988) The early work of John Logie Baird, *IEE Review*, **34**(8)
Chemand, A. (1991) HDTV stirs up French media, *World Broadcast News*, Intertec Publishing Corp., Kansas, p. 15, May
Chenard, S. The Soviet satellite industry: holding fast amidst difficult times, *Via Satellite*, **7**(3) p. 58
Childs, I. (1988) Putting you in the picture, *IEE Review*, **34**(7)
Clarke, P. *The Soviet Manned Space Programme*, Salamander Books, London, p. 10 and p. 176
Crutchfield, E. B. (1988) Broadcasting HDTV, *International Broadcasting Convention Technical Paper No. 293*, IEE, London, pp. 34–39
DePriest, G. L. and Schmidt, G. M. (1988) Advanced TV, a terrestrial perspective, *International Broadcasting Convention Technical Paper No. 293*, IEE, London, pp. 438–440
Dick, B. (1991) Building fibre optic transmission systems, *Broadcast Engineering*, November
Edwardson, S. M. (1988) DBS in the UK, home reception conditions, *International Broadcasting Convention Technical Paper No. 293*, IEE, London, pp. 200–203
Eglise, D. (1988) The UK DBS receiver, the requirements, *International Broadcasting Convention Technical Paper No. 293*, IEE, London, pp. 196–199
Forrest, J. R. (1987) Broadcasting to the future, *Electronics and Power*, **33**(11)
Free, L. R. (1988) An Antipodean view of HDTV, *International Broadcasting Convention Technical Paper No. 293*, IEE, London,

pp. 25–29
Gallois, P. (1987) Life expectancy of satellites, *Electronics and Power*, **33**(9)
Guedj, J. (1991) Fiber optic routing switchers, *Broadcast Engineering*, September
Hubbard, S. S. (1988) The US revolution in television through the use of satellites, *International Broadcasting Convention Technical Paper No. 293*, IEE, London, pp. 441–443
IEE News, September 1991, IEE, London, p. 16
Knight, J. F. (1991) Tracking inclined orbit satellites, *Journal of the Confederation of Aerial Industries*, **6**, 17–21
Kurtz, P. The future of fibre. *Broadcast Engineering*, pp. 74–92, May 1989
Lehinen, R. (1991) Integrating HDTV into NTSC, *Broadcast Engineering*, August
Lehinen, R. (1991) Comparing options in HDTV, *Broadcast Engineering*, August
Lilley, C. J. (1990) Intelsat's new generation, *IEE Review*, **36**(3)
Lopez, E. H. (1991) TV reaches islands by satellite, *World Broadcast News*, Intertec Publishing Corp., Kansas, April, p. 22
Luplow, W. C. (1990) Spectrum compatible high definition TV, *International Broadcasting Convention Technical Paper No. 327*, IEE, London, pp. 70–73
Matsushita, M. and Hasegawa, T. (1988) Experience in operating a direct broadcast satellite by NHK, *International Broadcasting Convention Technical Paper No. 293*, IEE, London, pp. 185–188
Millar, C. (1991) All light now; fibre optics, *IEE Review*, **37**(1)
Milman, J. A. F. (1988) Satellite services in the 1990s, *International Broadcasting Convention Technical Paper No. 293*, IEE, London, pp. 289–292
Moore, G. (1987) European spaceplanes: Hermes, Sanger or Hotol, *Electronics and Power*, **33**(4)
Mothersole, P. L. (1986) Developments in broadcasting technology, *Electronics and Power*, **32**(11)
Ninomiya, Y. and Okada, K. (1990) Television development and its application in HDTV, *International Broadcasting Convention Technical Paper No. 327*, IEE, London, pp. 6–9
Nishizawa, T., Enami, K., Tanaka, Y. and Kurita, T. (1988) HDTV and ADTV transmission systems, MUSE and its family, *International Broadcasting Convention Technical Paper No. 293*, IEE, London, pp. 37–40
Pascal, S. C. (1983) Launchers for communication systems, *Electronics and Power*, **29**
Potter, A. (1988) Britain's space programme, to be or not to be? *IEE Review*, **34**(4)
Pritchard, W. L. and Nelson, R. A. Launch vehicles for commercial satellites, *Via Satellite*, **7**(3) p. 94
Rankin, L. M. and Cooke, J. (1990) The business of advanced television, a Canadian perspective, *International Broadcasting Convention Technical Paper No. 327*, IEE, London, pp. 74–78
Robson, T. (1987) High definition; the technical challenge, *Electronics and Power*, **33**(2)
Rushin, D. (1989) Five decades of magnetic tape, *Broadcast Engineering*, May
Sabatier, J. (1990) HDTV, what do you mean? *International Broadcasting Convention Technical Paper No. 327*, IEE, London, pp. 408–411

Salkeld, B. (1988) The DBS system for the UK, *International Broadcasting Convention Technical Paper No. 293*, IEE, London, pp. 189–191
Sannikov, I. (1991) Satellite reception in USSR progresses, *World Broadcast News*, June
Saraga, P. S. (1989) Compatible high definition television, *Electronics and Communications*, **1**(1)
Sawanobori, S. (1991) NHK satellite system misfires, *World Broadcast News*, June
Smalling, E. (1988) Scrambling methods, *Broadcast Engineering*, May
Smalling, E. (1988) Up, up and away, *Broadcast Engineering*, August
Smalling, E. (1988) Getting into orbit, *Broadcast Engineering*, December
Smalling, E. (1989) A look back at 1988, sat and TV, *Broadcast Engineering*, January
Smalling, E. (1989) Spread spectrum technology, *Broadcast Engineering*, February
Smalling, E. (1989) Space robots at hand, *Broadcast Engineering*, April
Smalling, E. (1989) Fibre transmission seems to be the fix, *Broadcast Engineering*, March
Smalling, E. (1989) It's time for spring cleaning, *Broadcast Engineering*, May
Smalling, E. (1989) The business world turns to VSAT, *Broadcast Engineering*, June
Smalling, E. (1989) Sunspots affect communication, *Broadcast Engineering*, July
Smalling, E. (1989) The making of a space station, *Broadcast Engineering*, August
Smalling, E. (1989) Team goes to work in robot center, *Broadcast Engineering*, September
Smalling, E. (1989) Winter's coming: prepare now, *Broadcast Engineering*, October
Smalling, E. (1989) First bird launched 25 years ago, *Broadcast Engineering*, November
Suydam, M. (1990) Consortium narrows the race, *World Broadcast News*, March
Tamaka, Y., Kubota, K. and Iwade, Y. Transmission of HDTV signals by DBS and comms satellites, *International Broadcasting Convention Technical Paper No. 293*, IEE, London, p. 41
Tanaka, Y., Kubota, K. and Iwaadate, Y. (1988) Transmission of HDTV signals by DBS satellite and communication satellite, *International Broadcasting Convention Technical Paper No. 293*, IEE, London, pp. 41–44
Television and Video Engineer's Reference Book (1991) Butterworth-Heinemann, Oxford. Tables 28–4 to 6, Table 31–2, Figure 31–2
Thomson Tubes Electroniques to supply US DBS with TWTs, Press release, August 1991
Toth, A. G. Tsinberg, M. and Rhodes, C. W. (1988) Hierarchical NTSC compatible HDTV system, *International Broadcasting Convention Technical Paper No. 293*, IEE, London, pp. 30–33
Walker, G. (1991) HDTV: New aspects for the big picture, *World Broadcast News*, September
Walker, G. (1991) Europe: HDTV at Montreux, *World Broadcast*

News, September
Walker, G. (1991) A fiber diet prescribed, *World Broadcast News*, October
Walker, G. M. (Ed) (1991) Equipment survey, *World Broadcast News*, Intertec Publishing Corp., Kansas, pp. 45–50, September
Washington, D. (1989) Colour in flat panel CRTs, *Electronics and Communications*, **1**(1)
Whitaker, J. (1989) Milestones in the evolution of technology, *Broadcast Engineering*, May
Whitaker, J. (1989) Perspective on the industry, *Broadcast Engineering*, May
Whitaker, J. (1989) AM radio to the 21st century, *Broadcast Engineering*, May
Whitaker, J. (1989) The roots of AM broadcasting, *Broadcast Engineering*, September
Whitaker, J. (1989) Tape recording technology, *Broadcast Engineering*, November
Whitaker, J. (1989) Advancing technologies, *Broadcast Engineering*, December
Whitaker, J. (1989) Milestones in the evolution of technology, *Broadcast Engineering*, Intertec Publishing Corp., Kansas, June, p. 22
Williamson, M. (1989) Space technology at Le Bouget, *IEE Review*, **35**(7)
Williamson, M. (1989) The image from orbit, *IEE Review*, **35**(10)
Williamson, M. (1990) Space in the 1990s: promise and hope, *IEE Review*, **36**(1)
Williamson, M. (1990) Pidgeon and the eagle, *IEE Review*, **36**(1)
Williamson, M. (1990) Space insurance; a calculated risk, *IEE Review*, **36**(4)
Williamson, M. (1990) Ulysses, space odyssey over the sun, *IEE Review*, **36**(10)
Williamson, M. (1990) Comms are looking up, *IEE Review*, **36**(10)
Williamson, M. (1991) Space science over the horizon, *IEE Review*, **37**(1)
Windram, M. D. and Drury, G. M. (1988) Towards HDTV, *International Broadcasting Convention Technical Paper No. 293*, IEE, London, pp. 1–7
Wood, J. (1989) AM terrestrial broadcasting, past, present, and future, *IEE Review*, **35**(3)
Wood, J. (1989) AM terrestrial broadcasting, the technology of the super transmitter, *IEE Review*, **35**(4)
Wood, J. (1990) Speaking unto nations, international broadcasting, *IEE Review*, **36**(6)
Wood, J. (1991) Desert sounds; broadcasting to the Arab world, *IEE Review*, **37**(7)
Wood, J. (1991) Desert sounds, Letter, *IEE Review*, **37**(9)
Wood, J. (1991) Growth explosion in international broadcasting, *Telecommunications Policy*, February
Wood, J. (1992) *History of International Broadcasting*, Peter Peregrinus Ltd, (IEE)
Wood, J. (1992) *Satellite Communications and DBS Systems*, Butterworth-Heinemann, Oxford

Appendix

This appendix covers the latest satellite communications developments around the world. This global update covers every region of the world both on new satellite networks and on latest developments in the technology. Topics include:

- *Developments in satellites and launch vehicles (rockets).* Launch contracts, launch rockets; American, French, Chinese and Russian, payloads, manufacturers, satellites.
- *Satellites over Central and South America.* Mexico, Brazil and Argentina's national networks.
- *European satellite systems.* ASTRA and EUTELSAT: latest satellites coming into service.
- *North American direct broadcast satellites.* Rapid development of DBS in the US companies and satellites.
- *Global business satellite communication systems.* US commercial satellite companies; PanAmSat, Orion.
- *Asia-Pacific region satellite systems.* National networks of Indonesia, Malaysia, Thailand and Hong Kong.
- *Digital television and digital video compression.* Emergent technologies, companies and products.

1 Developments in satellites and launch rockets

Launch contracts

Hughes, the world's most successful builder of spacecraft and satellites for communications and broadcasting has signed a contract with McDonnell Douglas, America's builder of space launch vehicles to launch at least ten satellites beginning in 1998. This contract is between two of America's leading companies; Hughes Space and Communications International (HSCI) Inc., of Los Angeles, and McDonnell Douglas Corporation of St Louis, Missouri. If all options are exercised, the contract, signed in May 1995, could be worth approximately $US1.5 billion.

The rocket which is the subject of this contract is the new Intermediate-lift rocket designated Delta III now under development at McDonnell Douglas's plant in Huntingdon Beach, California. Delta III is the latest in a line of highly successful launch vehicles of the Delta class, Delta II having established a reliability of launch record which is second to none in the world. This latest rocket can carry nearly twice the payload of its stablemate and will easily be capable of launching Hughes' largest satellite, the three ton weight HS 601 into geostationary-transfer orbit (GTO).

Both companies anticipate the first launch of Delta III and its payload to take place during the first half of 1998, with the tenth launch by year 2002. Options are available to Hughes to extend launch agreements to the year 2005. This launch agreement is aimed at maintaining Hughes leadership in satellite communications. A quick launch is always hard to find as almost all the launch companies have full manifests and by firmly committing to ten launches. Hughes is assured of reliable and affordable space transportation for its satellites well into the twenty-first century. Equally, McDonnell Douglas benefits because it has secured an anchor customer and a guaranteed date for the maiden voyage of Delta III. According to the chairman of HSCI, this contract is a major step forward in bringing down the cost of transportation into space and it will bring a degree of stability to the space industry in America.

What Donald L. Cromer had in mind when he made that statement is something of great concern to all who work in the space industry, it is that ever-present risk of a launch failure. When a launch goes wrong, for whatever reason, it means that another spacecraft and another rocket have to be constructed and this introduces severe delays to the space program. There are no soft landings in the launch business. When a launch rocket blows up, or the satellite fails to get into orbit the result is nearly always catastrophic.

McDonnell Douglas first began manufacturing intercontinental ballistic missiles back in the 1960s and its range of commercial launch rockets are the descendants of those ICBMs, but Delta III is much more powerful and larger than its early ancestor. Delta II has a magnificent record of reliability and on this fact alone it is expected that its successor Delta III will have a considerable impact on the space market when it makes its first revenue-earning flight in 1998.

Space launch rockets

Today the term launch rocket has been largely replaced by the more correct description 'expendable launch vehicle' in the context of its role as a launcher of satellites while the term rocket is still used for military and defence. Since the end of the cold war there has been a significant reduction in expenditure on defence work and this has led to some shrinkage in America's space industry. However, the commercial business of launching satellites for communications and broadcasting is gathering increased momentum.

In 1994 Martin Marietta acquired the business of General Dynamics and along with it the Atlas rocket program. Thus Martin Marietta has two rockets: its Titan program and now the Atlas program (Atlas I, II, IIA and the IIAS). In 1994 the Lockheed Aircraft Company acquired an interest in the Russian 'Proton' rocket program marketed under a new joint company Lockheed Krunichev Energya (LKE). Even more surprising is that in 1995 Martin Marietta and Lockheed merged their space launch businesses.

The space launch business is now becoming highly competitive as a result of very attractive launch prices now being offered by China and Russia. The Great Wall Industry of China has experienced a couple of major failures in recent years, even so, with its powerful rockets it is highly attractive to the satellite companies who have the choice of the following launch vehicles.

Table A.1

Launch vehicle	*Manufacturing company*	*Country of origin*
Atlas program	Martin Marietta (acquired from General Dynamics)	USA
Titan program	Martin Marietta	USA
Delta program	McDonnell Douglas	USA
Ariane program	Arianespace	France
Long March program	Great Wall of China Industry	China
Proton rocket	Lockheed Krunichev Energya	Russia

Payload capacity

Increasingly, the trend is towards the employment of more powerful communications satellites and these have correspondingly heavier gross weights when fuelled up. In keeping with this heavier demand on the launch rockets these are being made more powerful. More powerful launch rockets also have the capacity of being able to launch dual payloads into geostationary transfer orbit (GTO). Some of the more powerful launch vehicles that have recently entered service or will do so in the next two years are given in Table A.2.

Table A.2

Launch vehicle	*Manufacturer*	*Country*	*Payload capacity*
Atlas 2AS	Martin Marietta	USA	3629 kg
Delta III	McDonnell Douglas	USA	3818 kg
Proton	LKE	Russia	5700 kg*
Ariane IV	Arianespace	French	4500 kg*
Ariane V	Arianespace	French	6800 kg

*Signifies these rockets are in service, others expected 1986–8.

The satellite manufacturers

A combination of rising production costs, necessary research and development and the cost of maintaining highly expensive environmental testing chambers added to a shrinking military space program has led to company mergers or takeovers in America and Europe. In the US, the number of major players has shrunk to three. First, Loral swallowed Ford Aerospace, Martin Marietta acquired GE Astro Space and in 1995 merged with Lockheed space industry.

In Europe there has been a similar contraction in the space industry. Finally, British Aerospace succeeded in selling off its space business to the French company Matra. Thus Europe is down to just two major companies: Matra Marconi and another French company, Aerospatiale.

2 Satellites over Central and South America

That stretch of landmass which extends from Mexico's border with the US and which reaches all the way down to the southernmost tip of Chile in South America is one defined by huge cities, small towns and sparsely populated villages often separated by mighty rivers, massive mountain ranges, valleys and impenitratable jungles. Within this huge landmass known as Central and South America there are twenty odd nations, a few of which have embarked upon a technological revolution that will propel this continent into the twenty-first century.

Mexico

Mexico, a country of 96 million people was the first Latin American nation to realize the potential that satellites offered to countries with harsh topographical features which present barriers to other forms of radio communication and television broadcasting. It was this realization which acted as the spur to Mexico becoming the first-ever nation to invest in a national satellite system. Fifteen years ago in 1980 Mexico planned its satellite system and, in 1982, it had two satellites from Hughes, type HS376 in service. These two spacecraft were the first step in a plan that would project Mexico as a major player in space communications by the twenty-first century.

In 1991 México placed its second contract with Hughes and two further satellites for its new Solidaridad network. Solidaridad 1 and 2 were launched in November 1993 and October 1994 respectively. These new satellites are considerably more powerful than the two of the Morelos network, and each satellite supports 18 C band and 16 Ku band transponders. This Solidaridad satellite network projects three satellite footprints over Mexico, the Caribbean region, Venezuela, Colombia, Peru, Chile, Argentina, Paraguay, Uruguay and Bolivia. The uplink for this satellite network is at Iztapalapa, just south of Mexico City.

Brazil

Mexico led the charge in the technological revolution to bring satellite communications to Latin America but Brazil was not far behind. Following in the wake of Mexico with its 96 million Spanish-speaking peoples, Brazil an even larger nation of 161 million Portuguese-speaking inhabitants moved towards satellite technology three years after Mexico and by 1985 had two satellites type HS376 in orbit over South America. For its second phase of the Brazilsat network it chose two further HS376 of the wide-bodied type. These two HS376W satellites were launched in August 1994 and in March 1995 respectively. Both were launched from Ariane IV rockets at the French Guiana launch site, Kourou.

This new satellite network called Brazilsat B was done in close cooperation between Hughes who supplied the satellites and the Brazilian Institute of Space Research, and like the Mexico satellite network it, too, has been engineered to take adjacent countries such as Argentina, Uruguay and Paraguay within its satellites footprints.

Argentina

Following the example set by Mexico and later Brazil, Argentina, the third most populated country in South America with its 33.75 million Spanish-speaking peoples, will become a satellite power by the year 1997. This will happen when its telecommunications satellite NAHUEL comes into service. As with other countries in the South American continent Argentina is a country characterized by mountains, rivers and jungles. It is also characterized by differences between its large cities and its rural regions, its major cities have cellular telephones, telephone systems and some cable TV networks and the capital city Buenos Aires will shortly have fibre optic networks. In the rural regions of Argentina the situation is vastly different, poor television reception is normal while telecommunications services are often inadequate to meet day-to-day needs. Television reception from terrestial transmitters suffers from propagation problems that are encountered in countries that have jungles and mountains.

A satellite system which would provide transmission services for telephone, data, video services and television is the obvious solution. With this in mind the Argentine Government granted a license to a European consortium of three companies; Deutsche Aerospace, Aerospatiale and Alenia Spazio to provide such a service. Its first satellite NAHUEN will be equipped with 18 Ku band transponders of 56 MHz bandwidth which will have sufficient capacity to provide 36 TV channels or 18,000 telephone channels or a combination of both services. It is expected that NAHUEN will eventually cater for a VSAT market.

NAHUEN will be launched in the near future and will be positioned in a geosynchronous orbit at 72 deg WL with an expected life in orbit of more than twelve years. A second satellite is scheduled for deployment by year 2009 or sooner. The footprint coverage of this first satellite is designed to cover Argentina and neighbouring countries Chile, Paraguay and Uruguay, thus effectively serving a total population of 55 million Spanish-speaking peoples.

3 European satellite television

Europe's two big providers of satellite television are different in character and different in ownership. Now Astra and Eutelsat are gearing up to be first in the race to introduce digital television programs to viewers in the European countries on a direct-to-home basis through DBS satellites. When Eutelsat was first formed it was not a DBS broadcaster because direct-to-home satellite broadcasting had not then been conceived. It was a public service broadcaster funded through a consortium of European nations and its principal role was to relay television programs from one European country to another.

Astra on the other hand is a private commercial broadcaster owned by Societé Européene des Satelllite (SES), a Luxembourg-based company. Astra and Eutelsat are now making considerable investment in new, higher-powered satellites which can deliver high quality TV programs to viewers.

Astra

The number of Astra-receiving TV households in Europe continues to gather momentum. At the end of 1994, of the 160 million TV households in the twenty-two different countries in Europe and within reach of Astra, more than 35% were able to receive one or more channels from Astra. Program deliveries include direct-to-home TV and to cable outlets. By the year 1998 it will have no less than eight high-power DBS satellites in service, six of which will be Hughes' most powerful satellite type HS 601. See Table A.3.

Table A.3

Satellite	*Type*	*Manufacturer*	*Launch rocket*	*Launch date*
1A	GE 3000	GE Astro	Ariane IV	December 1988
1B	GE 3000	GE Astro	Ariane IV	March 1991
1C	GE 3000	GE Astro	Ariane IV	May 1993
1D	HS 601	Hughes	Ariane IV	October 1994
1E	HS 601	Hughes	Ariane IV	November 1995
1F	HS 601	Hughes	Proton	Summer 1996
1G	HS 601	Hughes	NYD*	Summer 1997
1H	HS 601	Hughes	NYD	Summer 1998

*NYD = not yet determined.

Financial details are not known but the size of the financial investment in placing eight satellites in service may be roughly deduced from the average or typical cost of a satellite as $US100 million, to which has to be added the cost of the rocket launch and its insurance which for one satellite can add a further $US150 million to the cost.

In the past Astra has used Ariane IV rockets but the summer 1996 launch is said to be by a Russian Proton. This Russian rocket has achieved a good reliability record. SES has stated its intention to start up digital TV from satellite 1F onwards, but will continue to broadcast satellite services in analogue format as at present.

Eutelsat: the new hot birds

Over the most recent years Eutelsat has increased its penetration into the satellite television market and by 1994 its 11-F1 satellite was carrying sixteen television channels and twelve radio channels. In February 1995 the company in a statement said that out of the 68.4 million homes that watch satellite TV in Europe, an estimated 47 million receive channels from the 13 degrees EL orbital location. Now, Eutelsat is bringing into service its third and latest generation of high power satellites. These have been designated the term 'hot bird' because of their power. Hot Bird 1 has already been launched and Hot Bird 2 is scheduled to follow Hot Bird 1 by summer 1996. These new and powerful satellites will strengthen Eutelsat's position in Europe as a DBS broadcaster. The output power of the transponders will permit direct-to-home reception using 70 cm dish size antennas over an area stretching from Spain in the West, to the Ukraine, while giving the same performance over the British Isles where Astra is in strong competition with Eutelsat.

Eutelsat's new hot birds will be equipped to transmit analogue and digital TV within the same transponder thus guaranteeing true simulcast reception. This means that no viewer will be deprived of reception if that viewer has not got the digital decoder. At present these decoders are not available but are expected to be so by 1996.

Table A.4 *Eutelsat hot bird program*

Satellite	*Manufacturer*	*Transponders*	*Launch rocket*	*In service*
Hot Bird 1	Matra-Marconi	16 transponders of 55 watts	Ariane IV	March 1995
Hot Bird 2	Matra-Marconi	20 transponders of 118 watts	Ariane IV	September 1996
Hot Bird 3	Matra-Marconi	20 transponders of118 watts	Proton	March 1997

All satellites will be positioned at orbital slot 13 degrees WL.

A feature of Hot Bird 3 will be a steerable beam antenna connected to six of its twenty transponders, thus giving the capability of delivering a 49 dBW spot beam to any place visible from the satellite at 13 degrees EL.

4 North American direct broadcast satellites

Exactly as planned at 9.24 p.m. local time on 9 June 1995 an Ariane IV rocket roared skywards from the Guiana Space Launch Centre in French Guiana. Its payload was DBS-3, the third in a series of high power direct-to-home TV broadcast satellites. The launch was perfect and this satellite has now joined its stablemates. DBS-1, DBS-2 and now DBS-3 are co-located at the same orbital slot; 101 degrees West longtitude. DBS-1 was launched in December 1993 and in the process made history as America's first high power DBS satellite.

Though all three satellites are identical, the first was launched from an Ariane IV rocket, the second by an Atlas II rocket and the third was launched from an Ariane IV rocket – which just goes to show how the space launch business has become highly adaptive. All three DBS satellites have 4300 watts of transponder power which makes them the most powerful DBS satellites in service but the makers, Hughes, have an even higher powered satellite which is scheduled for launching in 1997. It will use gallium arsenide solar cells which will be capable of generating extra power.

When America does anything it is with style and promptness. It is not all that long ago when the US broadcast industry pondered over whether to introduce DBS to its viewers. Now it has three satellites in service and many others due to enter service. DBS-1 is shared by Hughes company, DirecTV and another company USSB. DBS-2 is used by DirecTV on an exclusive basis while DBS-3 is an in-orbit spare but also acts as back-up for DBS-1 supplying additional transponder capacity.

Altogether these three satellites have the capacity to deliver in excess of 200 DBS entertainment TV channels to its viewers who will need to be equipped with a DSS digital home receiving unit, manufactured by RCA/Thomson Consumer Electronics. Viewers need only an 18 inch dish size antenna anywhere in the US.

These DBS satellites operate in the BSS portion of the Ku band spectrum 12.2–12.7 GHz and use circular polarization. Each spacecraft has 16 TWT amplifiers of 120 watts output but which are also capable of being switched to 8 transponders of 240 watts output power. Depending on the configuration, the satellites can deliver between 48 to 53 dBW of radiated power over the Conus region (continental USA) and Southern Canada.

The operations control centre for these DBS satellites is at Hughes Space and Communications Headquarters at El Segundo, California. Telemetry and command terminals are at Castle Rock, Colorado and in Spring Creek, New York. The uplink terminal is designed to handle and transmit 216 simultaneous TV broadcast channels.

Hughes, with its DirecTV company, and United States Satellite Broadcasting (USSB) are not the only two DBS broadcasters. They have some major competitors planning to be in service by 1996. EchoStar, another DBS broadcaster, plans to have two high power satellites in service sometime in mid 1996. These satellites are being manufactured by Martin Marietta and are type series 7000. These are due to be launched from China's space launch center from Long March rockets type 2E.

A fourth major competitor is Tempo. It will use satellites from yet a third satellite manufacturer; Space Systems/Loral. These will be equipped with 32 Ku band transponders with output powers of 107 watts. The planned dates for launching these satellites is 1996 and it is thought the launch rockets will be the Russian Proton.

Thus by 1998 viewers in the whole of the United States and parts of Canada are expected to have a glittering choice from which to select from. DirecTV with its 200 channels, USSB with probably another 100 channels, EchoStar with its 100 channels, Tempo with its 200 channels and probably others all add up to a galaxy of choice for American viewers.

5 Global business satellite communications systems

PanAmSat, which made history when it launched its first communications satellite in the spring of 1988, was a success from the day it was launched. It offers specialized innovative satellite communication services to the institutions and business users, and through its infrastructure it provides end-to-end services, and users can access the satellite network with shared antennas and terrestrial links on customer premises. The network is highly suitable for broadcasters, pay-TV services, cable networks, program syndicates, news gathering and radio networks (see page 39–41).

The first satellite has served the Atlantic Ocean region since 1988 but such was the business potential of PanAmSat that by 1993 it had already begun to plan for a further three satellites. PanAmSat's first satellite PAS-1 was a GE Astro Series 3000 but for the three follow-on satellites; PAS-2, PAS-3 and PAS-4, the company selected the Hughes satellite type HS 601. PAS-2 was launched July 8 1994 from an Ariane IV rocket. The next satellite in the series, PAS-3 was lost in an unuccessful launch from an Ariane IV rocket in December 1994. Its replacement PAS-/R is under construction and scheduled for a space launch in December 1995 but it is likely that PAS-4 will be launched before the replacement PAS-3R.

In March 1995 it was announced that PanAmSat Corporation had placed another order with the Hughes Aircraft Company for an enhanced version of the HS-601. It is designated HS-601HP and will be the most powerful high power satellite by the time it enters service with PanAmSat Corporation in the first quarter of 1997. PAS-5 will have 24 high power transponders in both Ku band and C band, with 7 kilowatts of total power, almost twice the power of the standard HS-601. According to a statement from Hughes the HS-601HP features some new technologies in its design.

At the time of writing, this latest satellite is designated as a spare and its orbital position has not yet been defined by PanAmSat Corporation. The company has until 1999 to select a coverage area, launch date and the launch contractor.

The HS-601 series of high power spacecraft is a cube-shaped bus, of body stabilized design, with the solar panel wings extending from the north and south sides, and an antenna array. With solar wings unfolded and antennas deployed the satellite measures 86 feet from end-to-end and 23 feet across. At the commencement of its orbital life an HS-601 weighs approximately 3800 lbs. The flight-proven bipropellant propulsion system features an integral 490 newton (110 lbs/ft) Marquardt liquid apogee motor plus twelve 22 newton thrusters for stationkeeping.

Table A.5 *PanAmSat orbital locations*

Satellite	*Ocean region*	*Orbital location*
PAS-1	Atlantic	45 WL*
PAS-2	Pacific	192 WL
PAS-3	Atlantic	43 WL
PAS-4	Indian	292/288 WL

*45 WL means an orbital slot of 45 degrees, West longitude.

PanAmSat Broadcast services offer a wide range of facilities to create or expand domestic, regional and global program distribution making it possible to link urban centres on a national or international level and to reach those isolated regions within a large country. With satellites now covering all three ocean regions of the world, PanAmSat can broadcast television and video services throughout the world. It offers to customers the choice of full-time dedicated capacity on its satellites, long-term full-time lease or long-term part-time use. Effectively, PanAmSat is competing with Intelsat for satellite traffic but with the advantage that it can offer specialized services.

Orion Atlantic

Orion Atlantic is the newest and latest satellite communications company to enter the satcoms market to satisfy the ever-increasing demand for more satellite capacity in the transatlantic region. Throughout the history of transatlantic communications, right from the early days of cable telegraph, then longwave telegraphy, through to HF shortwave communications and then the satellite age it seems that the demand for more and faster communications with greater carrying capacity is insatiable.

PanAmSat Corporation was the first private company to enter the satellite communications business, and now the latest company to enter the field is Orion Atlantic. At the present time Orion has only one satellite. Where Orion differs is that it was formed from a business partnership of eight companies. But these are no ordinary companies, they are eight of the world's biggest multi-nationals drawn from America and Europe and they are all in the space and communications industry.

The partners are: British Aerospace, (UK), COM DEV (Canada), Kingston Communications (UK), Martin Marietta (US), Matra Hachette (France), Nissho Iwai (Japan), Orion Network Systems (US) and STET (Italy). The new company claims it is structured unlike any other telecommunications organization, that it can deliver true end-to-end control of network and one single point of accountability for sales, service and support.

In the beginning of the satellite communications age there was just the one provider, Intelsat. The rapidly growing demands for transatlantic communications and elsewhere in the world have led to the de-regulation of the satellite communications market.

Orion Atlantic is self-sufficient in both capital and expertise, an examination of the company profile satisfies this criteria. Orion Satellite Corporation is the general partner and management arm and will oversee the technical side of satellites and launch, with Orion Satellite Corp., being responsible for centralized marketing of satellite capacity. The stated objectives are to provide transatlantic and Pan-European satellite capacity services to radio and TV broadcasters, cable companies, multiple systems operators and telecommunications providers.

The (first) satellite Orion 1 was launched from Cape Canaveral on 29 November 1994 and is located at 37.5 WL. Its 34 transponder, all Ku-band satellite provides multiple spot beam coverage, and also a broad beam coverage of North America and Europe. Orion 1 was launched by a Martin Marietta launch vehicle, at Atlas IIA, on 29 November 1994, and is now correctly positioned in its alloted orbital position at 37.5 WL. Orion 1 is a three-axis type body stabilized satellite with a designed orbital life span of 10.5 years and the prime contractor for manufacture and delivery in orbit was Matra Marconi.

Orion 1 was specifically optimized for small diameter earth stations and digital services. It provides flexible communications options which include voice, audio, data distribution networks and transmissions in digital and analogue formats. Orion Atlantic has selected General Instrument's MPEG-2 digital video compression which can improve the transmission efficiency rate between 4:1 and

10:1. Analogue video rates for bandwidths between 27 and 36 MHz can cost upwards of $US2000 per hour between the US and Europe but the new digital technology combined with compression can drastically reduce this figure.

Proof of the fact that the north Atlantic route is the fastest growth sector of the world's communication markets, is that Orion Atlantic has in the short time since its satellite became operational has received some long-term contracts to carry communications for some of the big corporations in the United States and in European countries, Denmark and the Czech Republic.

As from 1 September 1995 the company placed full-time video signals on one of the transponders, transponder 12, to assist new customers to point and align their antennas towards the satellite. This video transmission is coming from Dallas Fort Worth Teleport in PAL 625/50.

Orion Atlantic is a private international partnership, with headquarters in Rockville, Maryland, USA.

Faxcast

This is a new method of delivering business documents or letters to a large number of subscribers. Faxcast, operated by the company with the same name, is a delivery system which one day may be the means of receiving a small newspaper. Faxcast is not a new technology but is an innovative idea which makes use of existing technology. It transmits its signals through the vertical blanking signals on a TV transmission, not visible to the viewer. The Faxcast broadcast which can be sent over a terrestrial link or by a satellite is then decoded by the Faxcast receiver and then printed out in the normal way. Its chief advantage over a facsimile message is that the same message can be sent out and received simultaneolusly by any number of subscribers with a significant decrease in transmission costs. After a period of trials around the world Faxcast is now fully operational. The company has concluded arrangements with satellite operators in the US, Europe, and in other parts of the world to carry Faxcast broadcasts. To what extent this technology replaces the use of dedicated VSAT terminals remains to be seen.

6 Asia-Pacific region satellite systems

The balance of economic power in the world is shifting to East, Central and South-East Asia and the Pacific rim region. Already many of those countries that were once regarded as poor, now have vibrant economies with economic growth figures of 8 to 11% and still rising. These nations have demonstrated an ability to surge ahead while many countries in Western Europe are experiencing the reverse, with fallings in GDP. Asia now satisfies much of the total world markets in consumer electronics.

From the beginning of satellite communications many of these emergent countries realized its potential and invested in this new technology. For other Asian countries satellite communications was not just a luxury but a necessity because of the geographical nature of the country. Indonesia, for example, is a fragmented archipelago of islands which stretch over 1500 miles down to the South Pacific. It was natural then that this nation should be the first in South-East Asia to invest in a satellite telecommunications network. No other form of radio communications could provide what was needed. In early 1970, Indonesia began planning of a satellite network, to be called the Palapa network, and in 1976 the first of its satellites was launched. By 1993 it had a network of operational satellites, three HS-376 models, and in 1993 it placed an order for three of the most powerful satellites in the world, the HS-601 type

Japan in North-East Asia was already an advanced nation when the satellite age was ushered in and acquired several satellites. This year the Space and Communications Corporation of Tokyo, placed a contract with Hughes Space and Communications Inc (HCSI) to construct a high power satellite. This new satellite, to be called Superbird C is scheduled for delivery in early 1997 and it will have a life in orbit of thirteen years.

This new satellite will have 24 active transponders which will carry television and business communications throughout the whole of Japan, to South and South East Asia and as far as Hawaii. The spacecraft will have some specially designed antennas using Hughes innovative shaped-beam technology. The launch contractor has not yet been chosen at the time of publication.

China and Hong Kong

In March 1995 the APT Satellite Company of Hong Kong ordered its new satellite. It will be designated Apstar 1A and will be almost identical to Apstar 1. Both have 24 transponders, all in the C band, and will carry television and other communications services. Coverage area of this new satellite will be somewhat larger than Apstar 1 and will take in China, Central and East Asia and as far west as all of the Indian continent. Apstar 1A will be an HS-376 type spacecraft and will be available for launch within eleven months from the contract award date.

China: the ill-fated APSTAR 2

On 18 January 1995 the Great Wall Industries of China launched a satellite for the APT Satellite Communications Company. The satellite was built by Hughes and it was launched from China's Space Center in Xichang in the Sichuan Province, China. The launch vehicle was a Long

March 2E Rocket. Apstar 2 was a failed mission, it was lost when the rocket blew up in the ascent stage. This was the second such incident with the Chinese rocket, the previous and similar incident took place in December 1992 with Optus B2 satellite.

Hong Kong
In 1995 Asia Satellite Communications Company (ASIASAT) launched its second satellite. This is of Martin Marietta manufacture, Type MM 7000. MM 7000 is a very powerful satellite weighing 3.5 tonnes at launch. It is equipped with both C band and Ku band transponders with output powers of 55 watts and 115 watts respectively. It was launched from China's space launch center on a Long March 3 rocket. This new satellite is positioned at orbital slot 100.5 EL, the same as its stablemate Asiasat I.

With these two satellites Asiasat I and II, the company will be able to serve the whole of the Chinese Peoples Republic, and the countries of South-East Asia, Hong Kong, Taiwan, Thailand, Malaysia, Vietnam, India, Pakistan, Russia, Japan, Nepal, Korea, Laos, Cambodia, Myanmar, Iran, Bangladesh and much of Eastern Europe. In all, Asiasat will cover fifty-three different countries in Central and East Asia; 3.3 billion people which represent 63% of the world's population.

Thailand
Following in the wake of other countries in South-East Asia, the Thai Government granted a thirty-year concession to Shinawatra Computor and Communications Company to develop the country's first satellite communications system. Two lightweight versions of Hughes' HS-376 satellite were ordered to be built and delivered in 24 and 28 months respectively. The system was named 'Thaicom' to symbolize the link between Thailand and modern communications. Both satellites are now in service, the last one, Thaicom 2 was placed in orbit aboard an Ariane IV rocket on 7 October 1994.

These two satellites enable the company to provide the capability for private VSAT networks, direct-to-home television, digital audio broadcasting, mobile radio telephone services and videoconferencing. Thaicom 1, the earlier satellite, was launched from an Arianespace rocket from French Guiana on 17 December 1993.

Malaysia
Malaysia is no newcomer to modern technology. It has one of the fastest growing economies in South-East Asia and is setting itself a program to become one of the leading industrialized nations in that region of the world within the next ten years. Its ecomomic growth has risen from 6.5% in 1989 to 8% in 1993. Malaysia is not without experience in satellite communications having leased time on the Intelsat network and also on the Palapa network owned in Indonesia. The Malaysian satellite system has two satellites,

both are of the HS-376 type. The network is named Measat. Measat 1 is scheduled for launching in December 1995. Earlier in 1995 the company authorized Hughes Space and Communications Inc. to proceed with the manufacture of a second satellite; Measat 2.

Satellites Measat 1 and 2 are type HS-376 spacecraft. These satellites will enable Malaysia to provide a variety of services ranging from telephone, television, data transmission and business networks. The footprints of these satellites will cover India, Singapore, the Philippines, Thailand, Brunei, Indonesia, Vietnam, Kampuchea, Laos, Myanmar, Taiwan, Hong Kong and South China.

The Philippines

Of all the countries in South-East Asia the Philippines is the country which would benefit most from having a satellite system. This is because the country has been fragmented by nature. Its 7100 islands that make up the country make television impossible for many of its peoples. Terrestrial transmission systems cannot provide a solution but satellites can. In April 1991 the government began planning such a network and it would lease air-time on the Palapa satellite system owned by its neighbouring country Indonesia.

The final solution for the Philippines has to be a comprehensive satellite network owned by the country and which will provide telephone, data and television services to its 70 million people.

7 Digital transmission and digital video compression

Within the short space of time since the first edition of this book was written transmission technology has moved along at a breathless pace and has brought about some dramatic developments particularly in the field of television broadcasting. A decade ago it was the aim of the television industry to bring about a common standard and thereby eliminate the three present standards in use throughout the world: NTSC, PAL and Secam.

Thus Europe developed the D-MAC system and Japan developed its MUSE system. It seemed at the time that the world would adopt both these standards. But digital transmission was just around the corner under development in the United States.

D-MAC and MUSE

For all practical purposes both of these systems can now be considered as obsolete because they do not have a future. Both systems are still in use to a limited extent, some French broadcasters are still using D-MAC, and MUSE is in use in Japan but almost certainly nowhere else in the world. Both systems were excellent and could deliver good picture quality. The fact remained that they were analogue based, and doomed by that fact.

The satellite industry became aware of the advantages of digital video about a decade ago and its application to certain types of traffic; video conferences and business data began. Digital video offered the advantage that it can be compressed, so making it possible to carry several transmisions in the bandwidth of just one analogue TV transmission. DVC brings considerable advantages to the satellite industry. PanAmSat and some other broadcasters have been using DVC on many of the circuits to hotels and SMATV for a few years. However with satellite DBS the progress has not been as swift, the first all-digital direct broadcast by satellites will not commence until 1997.

There are no more technical problems to be overcome before digital compressed television is available to the general public, the only problem is the cost to the viewer. To receive a digital compressed signal a decompression unit is needed by the viewer and at this time no manufacturer has yet started mass production. As with every consumer item of electronics, mass production will lower production cost. However, a further problem is compatibility between different manufacturers of DVC systems. Although all the versions produced by six manufacturers are designed to meet the standard laid down by the Motion Picture Expert Group (MPEG) there is little or no compatibility which means that a set top DVC unit will not work on any other system.

In theory and in practice you cannot have compression without suffering something in picture quality and even if a viewer cannot detect any degradation in the picture quality it does not mean that there has not been any. The idea of compression is to take away redundant frames and thereby save on transmission bandwidth. Interestingly few programmers are concerned about reduction in picture quality. This is because viewers are primarily concerned with quality of programs and not the quality of transmission. The fact is that the introduction of digital video compression offers more to the satellite company than the end-customer. DVC is a technology to be reckoned with, and it is here to stay. The gain to the satellite operator or the transmission authority is quite considerable, for example sending six compressed TV channels down a bandwidth normally required for one analogue channel is almost like a license to print money because it reduces transmission costs.

Whether the introduction of digital TV offers the same advantage to a consumer is beside the point, the technology is attractive. Already more than 10% of all satellite communications uses digital compression while in specialized sectors like business satellite, distance learning and downlinking to hotel chains of pay-TV the percentage is much higher (around 50%).

As the technology becomes more popular with transmission authorities and end users, digitally compressed television will mature to a stage where there is compatibility between the different manufacturers products. At the time of preparing this appendix there are at least six manufacturers and probably more. These are given in Table A.6.

Table A.6

Company	*Country*	*Product*
Scientific Atlanta	USA	MPEG 1 and MPEG 2 standards
NTL	UK	System 2000 and 3000, fully meet all MPEG standards
General Instruments	USA	MPEG DIGital Compression System
Compression Labs Inc.	USA	SpectrumSaver
TV/COM	USA	Compression NetWORKS. Meet MPEG standards
Tadiran	Israel	CODICO. Meets MPEG 2 standards

As the market grows from professional users to the vast domestic consumer market it is certain that many other electronics manufacturers will move into the market. In this context one manufacturer well geared up for both professional and domestic markets is NTL of Winchester, UK. NTL video compression systems from this company are now in use throughout the world. Its System 3000 successor to the System 2000, is a complete multichannel video compression system for professional and direct-to-home TV. NTL has sold its system 3000 to APTV the video arm of Associated Press. Another of its major customers is the Shinawatra Group of Thailand. It initiated a direct-to-home digital TV service this year.

The US company General Instruments Corporation, a world leader in broadband transmission systems has recently signed some major contracts around the world for the supply of its digital compression systems. The latest version of this, its MPEG DIGital Compression is designed so as to permit widespread applications. One of its latest customers for this technology is CCTV, China Central Television of the Peoples Republic of China. CCTV intends to install GI's latest digital compression system on China's satellite ChinaSat 5, and also on its leased capacity on AsiaSat's newest satellite AsiaSat 2.

General Instrument Corporation is one of the founder members of the US 'grand alliance' of manufacturers of digital technology. Its objective is to share technologies with a view to better standardization and to produce a US standard for high definition television (HDTV). When that happens the world will be on the way towards establishing a common standard for HDTV

Index